NORTH AMERICAN FJ-3/3M FU

Above, the 8th FJ-3 built, BuNo 135801, with 200 gal. drop tanks, slatted wings, folded wings and deployed speed brakes. Intake lips were polished metal. (Larry Smalley)

LINEAGE: It was 1944, and a new age in aviation was dawning in the European skies, the jet age. As more German jets appeared in combat, it became a foregone conclusion that Japan would be producing them too. These facts prompted the Navy to issue a requirement for carrier based jet fighters. In late 1944, BuAer sent requests to several manufacturers for a jet fighter proposal. Up to that point only one all-jet fighter contract had been issued. It was for the twin-jet McDonnell XFD-1 (FH-1) Phantom (Naval Fighters #3). From the proposals submitted, the BuAer technical desk chose three designs to be developed. These were the McDonnell XFD-2 (F2H) Banshee (Naval Fighters #2 & #73), the Chance Vought XF6U Pirate (Naval Fighters #9) and the North American FJ-1 Fury (Naval Fighters #7).

Since these decisions were taking place in a wartime environment, the Navy could not rely on one airframe manufacturer or for that matter one engine manufacturer to produce a capable jet to counter the possible Japanese threat. Therefore , the F2H was designed around two of the Westinghouse J34 turbojets, the F6U around one Westinghouse J34, the FH-1 around two Westinghouse 19-XB turbojets and the FJ-1 around one General Electric TG-180 (Allison J35).

Since everyone was new to the jet business, there was considerable debate on how jet engines should be removed from airframes. North American approached the problem by making the entire upper fuselage from the cockpit aft to the fin removable. Vought on its F7U-1 lowered the engine out of the bottom of the fuselage (Naval Fighters #6). Lockheed's P/F-80 used the system of "breaking" the fuselage aft of the wing. This method involved removing the tail section and pulling the engine out the rear. This became the accepted method of removal and was used on the FJ-1's counterpart, the F-86, and on the rest of the Fury series. Even Vought went to this system by the time the F7U-1 grew into the F7U-3 Cutlass.

The initial contract for three XFJ-1 aircraft (NA-135) was signed on 1 January 1945. The Army's XP-86 was being developed concurrently with the XFJ-1. Originally, the planes were virtually identical. However, early in the design phase the Army decided to move the dive brakes from the wings to the fuselage. This change allowed the Army to install wing tanks and thus enabled the removal of the belly saddle tanks which provided the basis for a much cleaner and efficient fuselage design. The delay caused by the redesign of the XP-86 allowed for the opportunity to examine German reports on the benefit of swept wing technology. The Army decided to test the concept on the XP-86 and North American offered the design change to the Navy too. However, the Navy turned the offer down. The Navy was unsure as to whether jets could ever operate from carriers and did not want to concern itself with the added uncertainties of swept-wing performance.

First flight of the XFJ-1 was conducted in September 1946. After manufacturer testing was completed, the three prototypes (BuNos 39053-39055) were handed over to the Navy in September 1947. Delivery of production FJ-1s commenced on 5

Above, the first of the many, the FJ-1 Fury, sire of the F-86 series and Navy FJ-2, FJ-3/3M and FJ-4/4B. FJ-1 BuNo 120368 in Naval Air Reserve markings from NAS Oakland, CA. (USN) Below, the first XFJ-2, BuNo 133754, is tenstioned for catapult take-off tests at NAF Inyokern, CA. The prototype XFJ-2s were essentially navalized F-86Es. (NAA) Bottom, the third XFJ-2, BuNo 133756, in flight. (Ginter collection)

October 1947 when BuNo 120342 was accepted by the Naval Air Test Center (NATC). The FJ-1s entered service with VF-5A, redesignated VF-51, in August 1948 and along with VF-17s FH-1s demonstrated the feasibility of jet carrier operations. In May 1949 the squadron re-equipped with the F9F-3 Panther and the FJ-1s were transferred to the Naval Reserves.

Although the FJ-1 proved valuable in developing jet procedures aboard carriers, its performance was inadequate when compared with the XP-86 which flew ten months after the FJ-1. Because of the successful development of the swept-wing F-86, North American initiated project NA-181 for a navalized F-86 on 30 January 1951. The proposal was tendered to the Navy on 6 February 1951 and accepted on 10 February. Contract approval came on 29 May and mockup inspection was conducted from the 26th through the 28th of June 1951. The former skipper of the FJ-1 squadron VF-51, CDR Aurand, became the project officer.

Three pre-production prototypes of the XFJ-2 (Naval Fighters #10) were ordered by the Navy. Two aircraft, BuNos 133754 and 133755, designated NA-179 were ordered on 8 March. The third, BuNo 133756, designated NA-181 was ordered on 19 March 1951. These three aircraft were built at the Los Angeles plant, while the production aircraft were to be built at North American's Columbus, Ohio, factory.

133754 and 133755 were essentially F-86Es that were modified for carrier operations. The modifications included a V-frame arresting hook, catapult hooks and a lengthened nose gear. These two aircraft were also completed without armament.

The third XFJ-2 prototype, BuNo 133756, whose only modification was the installation of four 20mm cannons in place of the six .50 cal machine guns used on the FJ-1 and F-86, was the first to fly on 27 December 1951. It was designated XFJ-2B with the "B" denoting an armament modification and was sent to NAF Inyokern for weapons testing. As a result of the testing, production FJ-2s incorporated four 20mm guns with 600 rounds, a Mk.16 mod 2 gunsight and a AN/APG-30 radar.

The first XFJ-2 flew on 14 February 1952 and conducted initial catapult trials at NAF Inyokern before transfer along with the second XFJ-2 to the Naval Air Test Center. Carrier trials were conducted in December 1952 aboard the USS Coral Sea (CVB-43).

The 200 production FJ-2s were built with GE J47-GE-2 engines of 6,000 lbs. thrust, and as such were basically equivalent to the slatted-wing F-86F. Operationally, the FJ-2 was assigned to six Marine fighter squadrons from 1954 through 1956.

THE FJ-3 FURY: Design of a new Fury version, the FJ-3, began on 3 March 1952 and was designated NA-194 by North American. Designed around the new license-built Wright J65-W-2 engine of 7,800 lbs. thrust, 389 (BuNos 135774 through 136162) aircraft were ordered on 18 April 1952. A second contract for 214 FJ-3s (BuNos 141364-141443) designated NA-215 by North American was placed on 15 March 1954. 145 of these were later cancelled followed by 80 aircraft being added for a total of 149 completed. The last FJ-3 built was accepted by the Navy in August 1956.

Modification of the 5th FJ-2, BuNo 131931, under North American project NA-196 would result in the prototype FJ-3. It was fitted with a J65-W-2 engine and first flew on 3 July 1953. The intake on all production FJ-3s would be enlarged to accomdate the increased airflow needs of the larger engine.

Production FJ-3 Furies were fitted with J65-W-4 engines of 7,650 lbs. thrust and were initially built with the slatted wing of the FJ-2. This was replaced by a more cambered 6-3 wet wing holding an additional 124 gallons of fuel and possessing 14 sq ft more wing area. The wing also fea-

Above, prototype FJ-3 BuNo 131931 taxis after its first flight on 3 July 1953. (NAA)

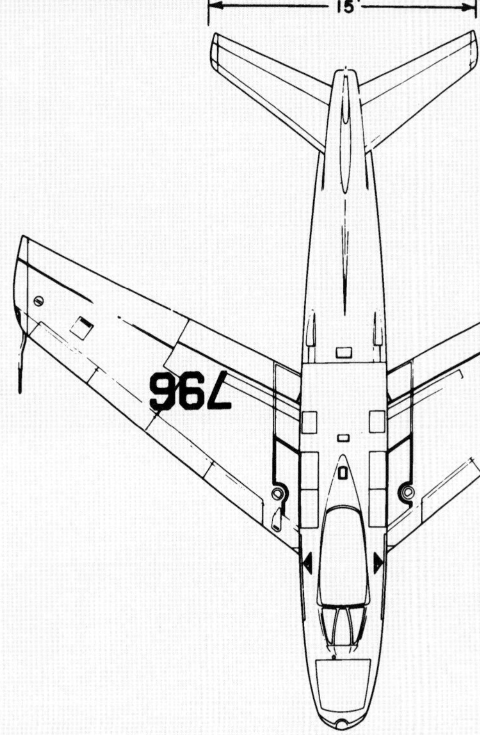

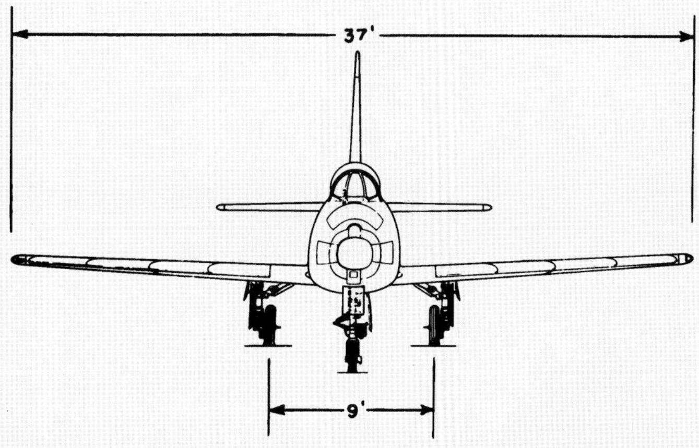

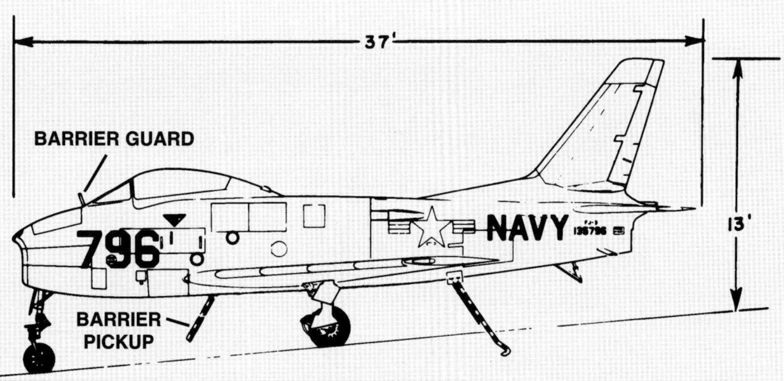

tured a small wing fence at about 65° span that wrapped around the leading edge and beneath the wing. Starting with BuNo 136118 the wings were equipped with four wing stations, provisions for inflight refueling and the Sidewinder air-to-air missiles. The Sidewinder-capable aircraft were designated FJ-3Ms.

On 11 December 1953, the first production, FJ-3 BuNo 135774, was rolled out at Columbus and was piloted by William Ingram on its first flight. 24 FJ-3s had been accepted by July 1954 and the Fleet Introduction Program (FIP) began in August at the Naval Air Test Center. The program was conducted by pilots from VC-3 and VF-173 and completed in four weeks with 703 flight hours. On completion of the FIP program, VC-3 and VF-173 became the first Navy squadrons to equip with the type.

Starting in 1957, a large number of aircraft were modified into Regulas and drone director aircraft designated FJ-3D (Regulas I) and FJ-3D2 Regulas II). Some of these aircraft were still operating after 1962 when they were redesignated DF-1C and DF-1D respectively.

Top to bottom: FJ-3 BuNo 135777 in flight. (USN) FJ-3 BuNo 136028 in flight. (USN) The 20,000th fighter produced by Noth American was an FJ-3. (NAA) FJ-3 BuNo 136110 at Edwards AFB in May 1957. Gear and leading edges were natural metal. (Bowers)

FJ-3 LEADING PARTICULARS

Span	37.12 ft
With Folded Wings	22.50 ft
Length	37.55 ft
Height	13.67 ft
With Folded Wings	15.76 ft
Design Gross Weight	16,482 lbs
Max Gross Weight	23,883 lbs
Wings:	
Chord Fus. Centerline	10.85 ft
Chord at Tip	5.64 ft
Incidence at Root	-0.22°
Incidence at Tip	-2.25°
Dihedral	3°
Sweepback	35.7°
Aspect Ratio	4.559
Horizontal Stabilizer:	
Span	15.08 ft
Chord Fuselage B/P 1.5	4.32 ft
Chord at Tip	1.95 ft
Incidence	6° up 10° down
Vertical Stabilizer:	
Height	9.15 ft
Chord at Root	6.38 ft
Chord at Tip	2.31 ft
Fuselage Width	5.00 ft
Fuselage Height	6.88 ft
Fuselage Length	35.61 ft

Top to bottom: FJ-3M BuNo 136153 on display at NAS Moffett Field on 17 May 1958. Intake and outer wing panels were red. (W.T. Larkins) FJ-3D2 BuNo 135797 drone controller at NAS Alameda. Aircraft had an engine grey fuselage with yellow and red wing and tail. (Smalley) Four FJ-3s BuNos 141364, 141375, 141377 and 141379, in formation. (NAA) FJ-3 BuNo 136135 in flight. (NAA)

PERFORMANCE SUMMARY

TAKE-OFF LOADING CONDITION		(1) Fighter
TAKE-OFF WEIGHT	lb.	19,360
Fuel (Gasoline)	lb.	2,610/2,400
Payload	lb.	- - -
Wing loading	lb./sq.ft.	67.2
Stall speed - power-off	kn.	111.5
Take-off run at S.L. - calm	ft.	2,050
Take-off run at S.L. 25 kn. wind	ft.	1,380
Take-off to clear 50 ft. - calm	ft.	- - -
Max. speed/altitude (A)	kn./ft.	533/11,000
Rate of climb at S.L. (B)	fpm	6,750
Time: S.L. to 20,000 ft. (B)	min.	3.9
Time: S.L. to 30,000 ft. (B)	min.	7.3
Service ceiling (100 fpm) (B)	ft.	43,200
Combat range	n.mi.	995
Average cruising speed	kn.	470
Cruising altitude(s)	ft.	41,300/46,200
Combat radius	n.mi.	* 320
Average cruising speed	kn.	470
Mission time	hrs.	1.7
COMBAT LOADING CONDITION		**(2) Clean**
COMBAT WEIGHT	lb.	16,600
Engine power		Military
Fuel	lb.	2,610
Combat speed/combat altitude	kn./ft.	530/35,000
Rate of climb/combat altitude	fpm/ft.	3,030/35,000
Combat ceiling (500 fpm)	ft.	47,000
Rate of climb at S.L.	fpm	9,400
Max. speed at S.L.	kn.	599
Max. speed/altitude	kn./ft.	599/S.L.
LANDING WEIGHT	lb.	14,833
Fuel	lb.	842
Stall speed - power-off	kn.	97.4
Stall speed - with approach power	kn.	94.4

Areas:

Wing	**302.32 sq ft**
Ailerons each	**16.36 sq ft**
Flaps both	**32.51 sq ft**
Horizontal Stabilizer	**47.18 sq ft**
Elevators each	**5.57 sq ft**
Vertical Stabilizer	**32.19 sq ft**
Rudder and Tab	**5.26 sq ft**
Rudder Trim Tab	**0.58 sq ft**

BuNos 139230-139278 and 141364-141443 had a splitter-type rudder with a trim tab of 0.93 sq ft

FJ-3/3M aircraft were produced with block numbers in lower case letters "a" through "i" applied after the BuNos.

These were:

135774-135786	a
135787-135812	b
135813-135852	c
135853-135912	d
135913-135982	e
135983-136072	f
136073-136162	g
139210-139278	h
141264-141443	i

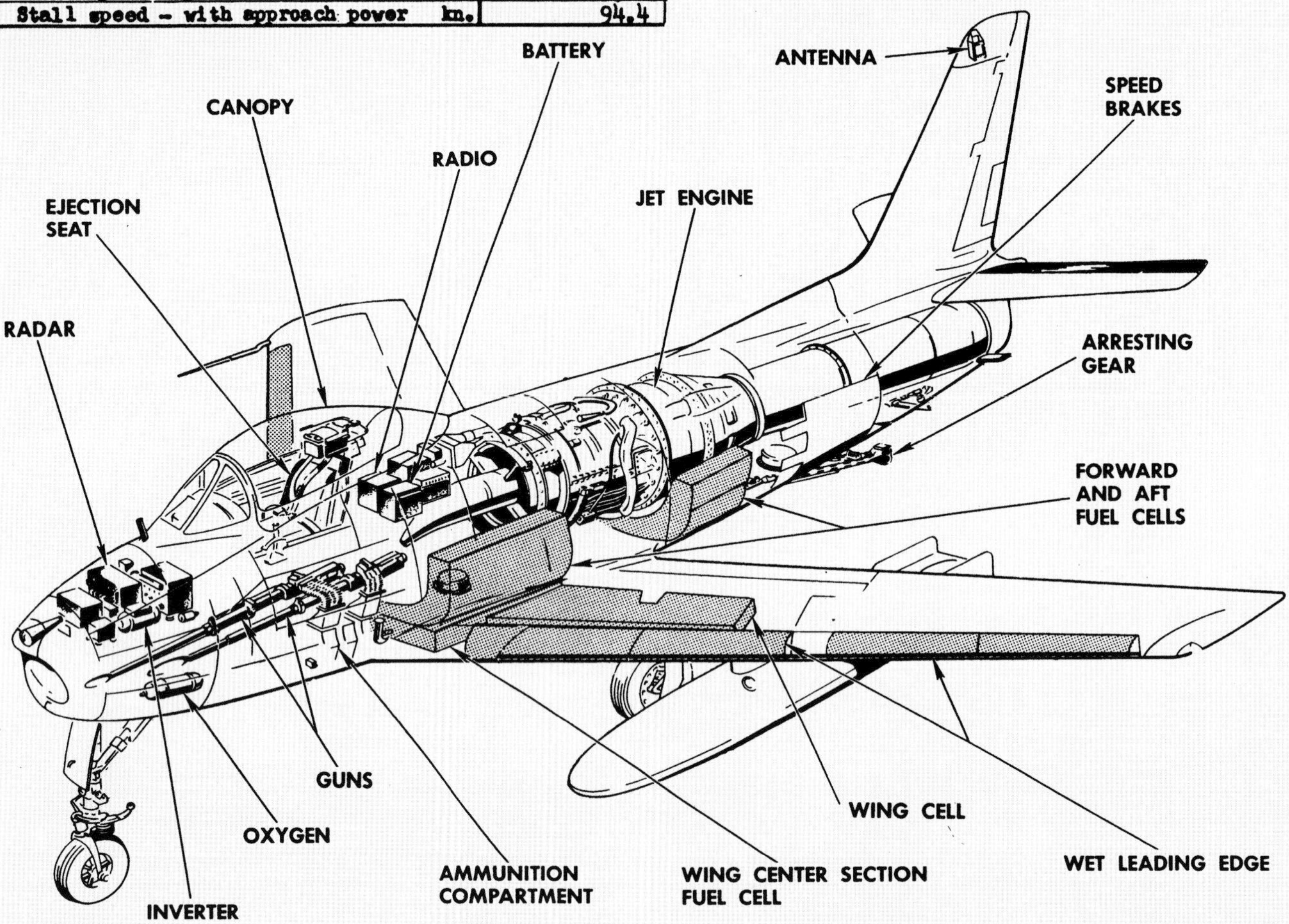

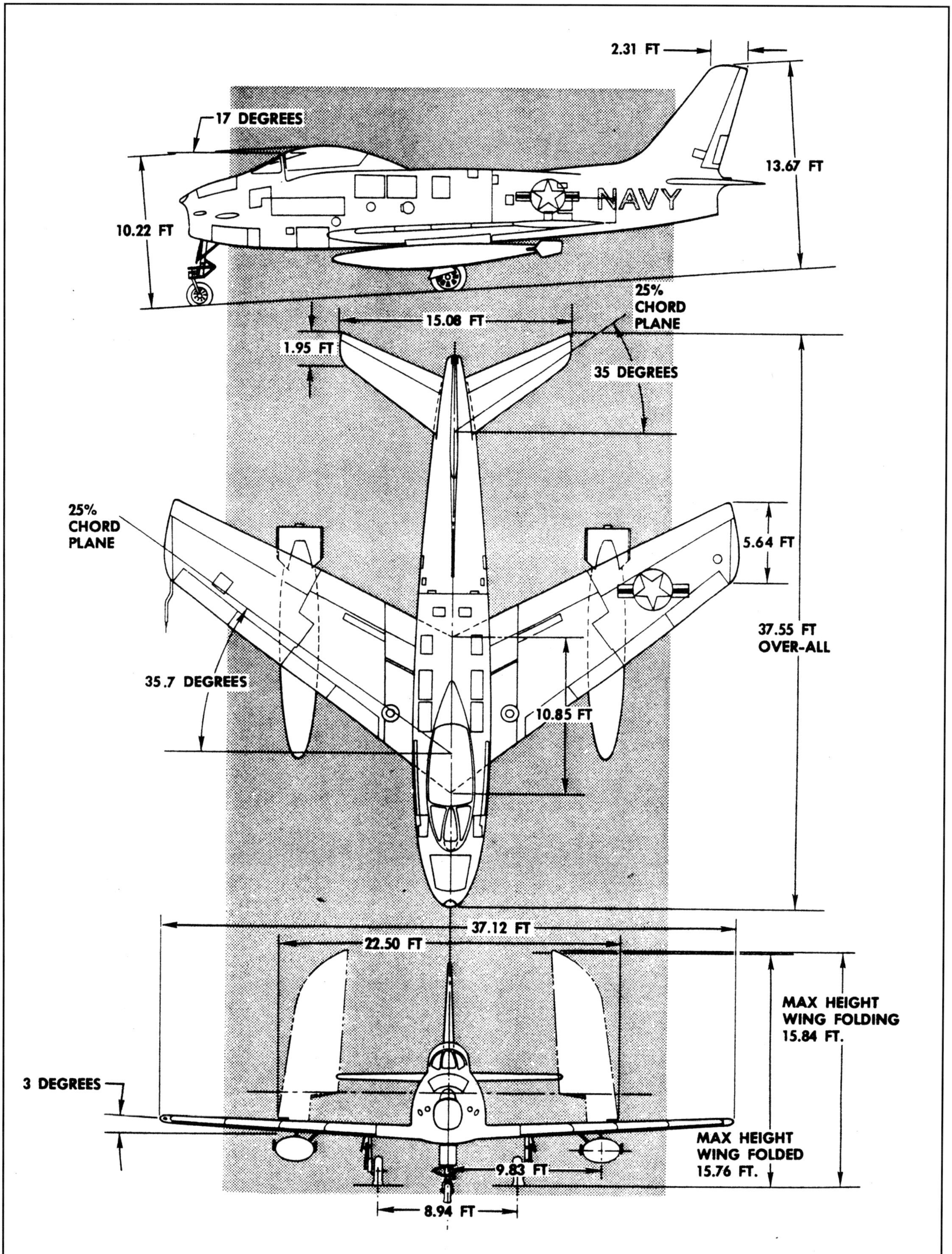

2.31 FT
17 DEGREES
13.67 FT
NAVY
10.22 FT
25% CHORD PLANE
15.08 FT
1.95 FT
35 DEGREES
25% CHORD PLANE
5.64 FT
37.55 FT OVER-ALL
35.7 DEGREES
10.85 FT
37.12 FT
22.50 FT
MAX HEIGHT WING FOLDING 15.84 FT.
3 DEGREES
MAX HEIGHT WING FOLDED 15.76 FT.
9.83 FT
8.94 FT

FJ-3 STATIONS DIAGRAM

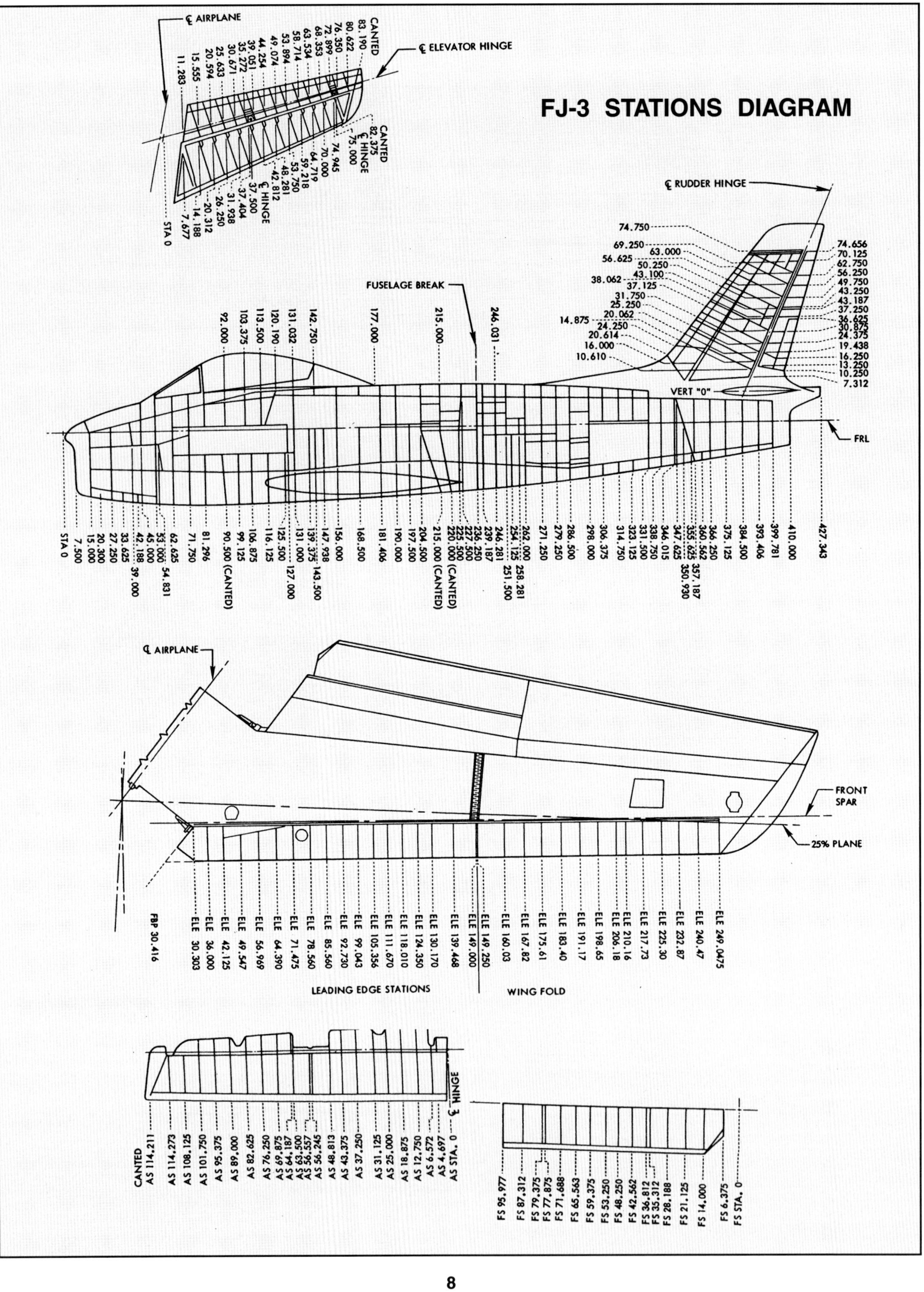

WRIGHT J65-W-4 SAPPHIRE ENGINE

The Wright license-built J65-W-4 engine had a sea level rating of 7,700 pounds thrust at 8,300 rpm. The 13-stage compressor section was fed air through a long nose intake duct before dumping into the annular combustion chamber for mixing with fuel. The exhaust then exited through a long exhaust pipe.

The compressor section was comprised seven low-pressure stages followed by six high-pressure stages. The first three rows of blades were steel, the next four were aluminium alloy and the last six were stainless steel.

The combustion chamber was of the single annular type with inner and outer stainless steel liners between which combustion took place. Thirty-six Armstrong Siddeley type fuel/air vaporisers were used. Secondary air entered the chamber through thirty-six air distribution cups to mix with combustion flames near the vaporizer tubes.

The turbine section was a two-stage type with 110 buckets in the first stage and 75 in the second stage.

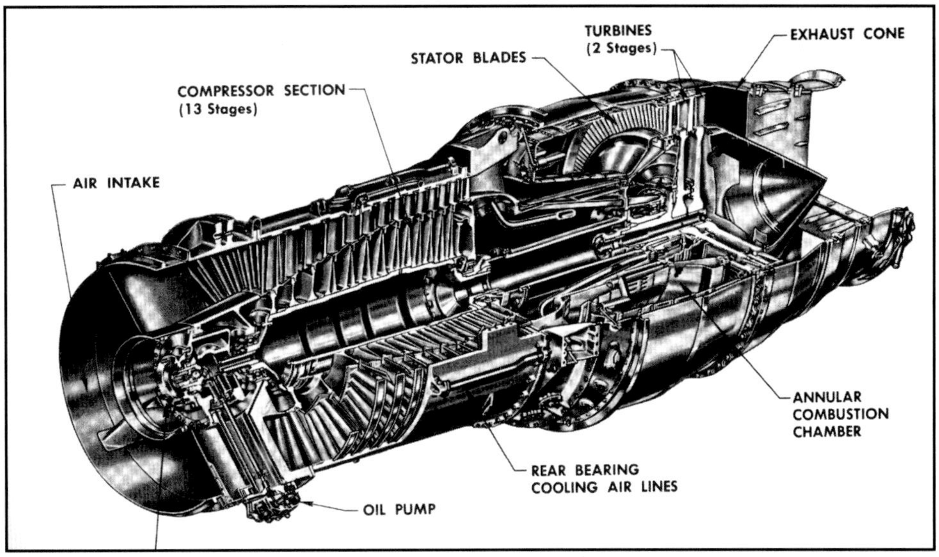

WRIGHT J65-W-4 SAPPHIRE ENGINE

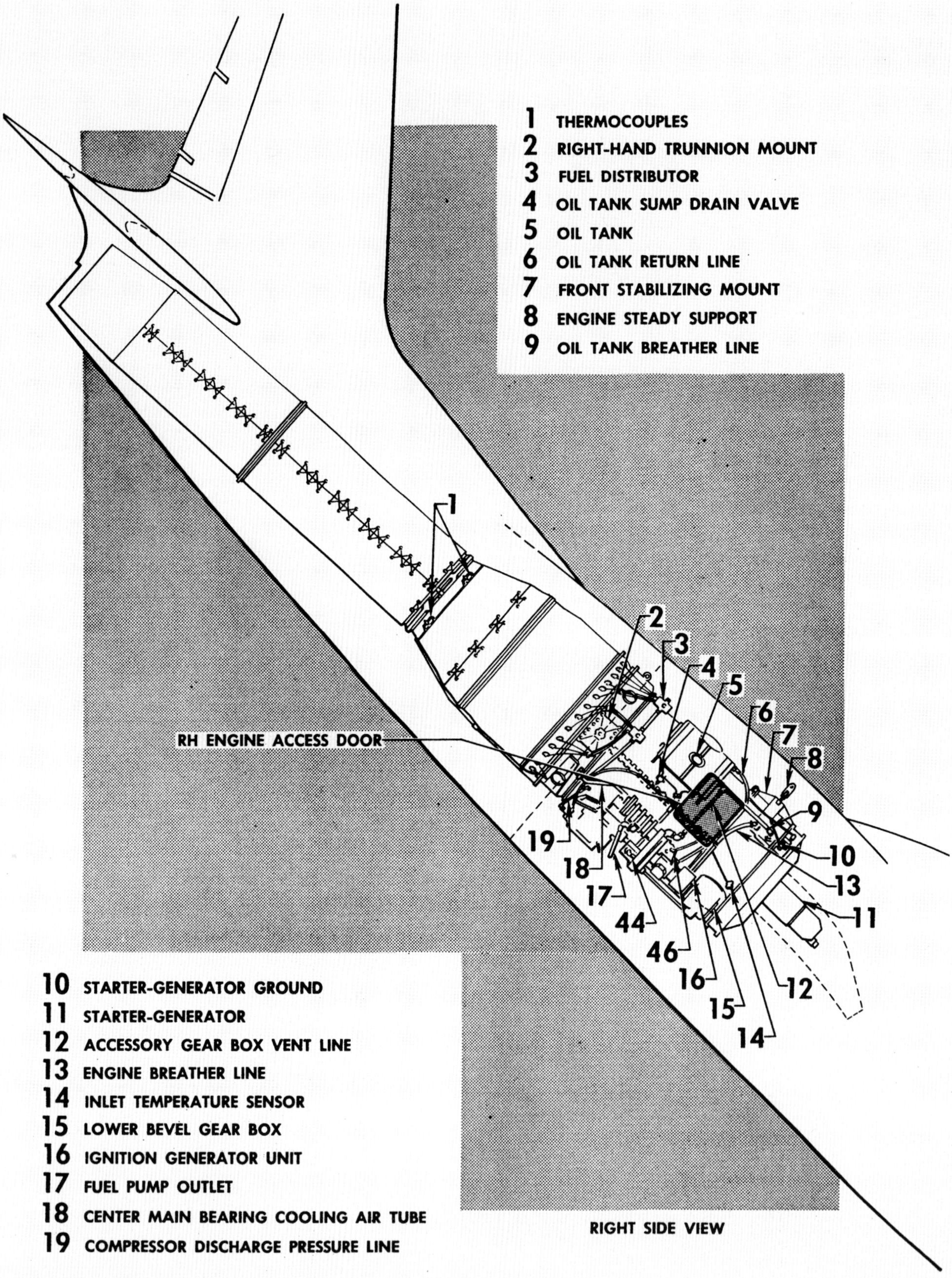

1 THERMOCOUPLES
2 RIGHT-HAND TRUNNION MOUNT
3 FUEL DISTRIBUTOR
4 OIL TANK SUMP DRAIN VALVE
5 OIL TANK
6 OIL TANK RETURN LINE
7 FRONT STABILIZING MOUNT
8 ENGINE STEADY SUPPORT
9 OIL TANK BREATHER LINE

10 STARTER-GENERATOR GROUND
11 STARTER-GENERATOR
12 ACCESSORY GEAR BOX VENT LINE
13 ENGINE BREATHER LINE
14 INLET TEMPERATURE SENSOR
15 LOWER BEVEL GEAR BOX
16 IGNITION GENERATOR UNIT
17 FUEL PUMP OUTLET
18 CENTER MAIN BEARING COOLING AIR TUBE
19 COMPRESSOR DISCHARGE PRESSURE LINE

RIGHT SIDE VIEW

WRIGHT J65-W-4 SAPPHIRE ENGINE

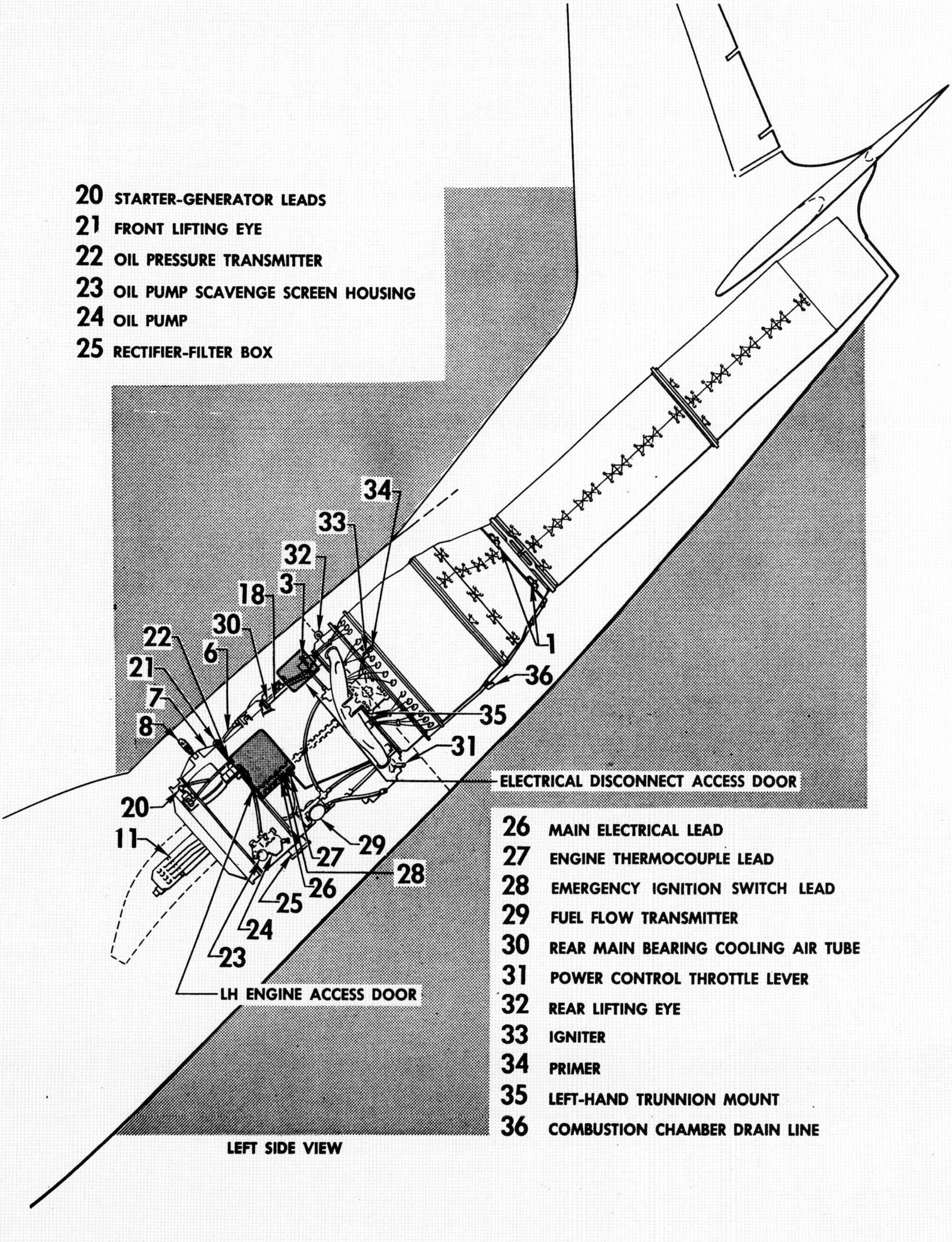

20 STARTER-GENERATOR LEADS
21 FRONT LIFTING EYE
22 OIL PRESSURE TRANSMITTER
23 OIL PUMP SCAVENGE SCREEN HOUSING
24 OIL PUMP
25 RECTIFIER-FILTER BOX

26 MAIN ELECTRICAL LEAD
27 ENGINE THERMOCOUPLE LEAD
28 EMERGENCY IGNITION SWITCH LEAD
29 FUEL FLOW TRANSMITTER
30 REAR MAIN BEARING COOLING AIR TUBE
31 POWER CONTROL THROTTLE LEVER
32 REAR LIFTING EYE
33 IGNITER
34 PRIMER
35 LEFT-HAND TRUNNION MOUNT
36 COMBUSTION CHAMBER DRAIN LINE

LEFT SIDE VIEW

WRIGHT J65-W-4 SAPPHIRE ENGINE

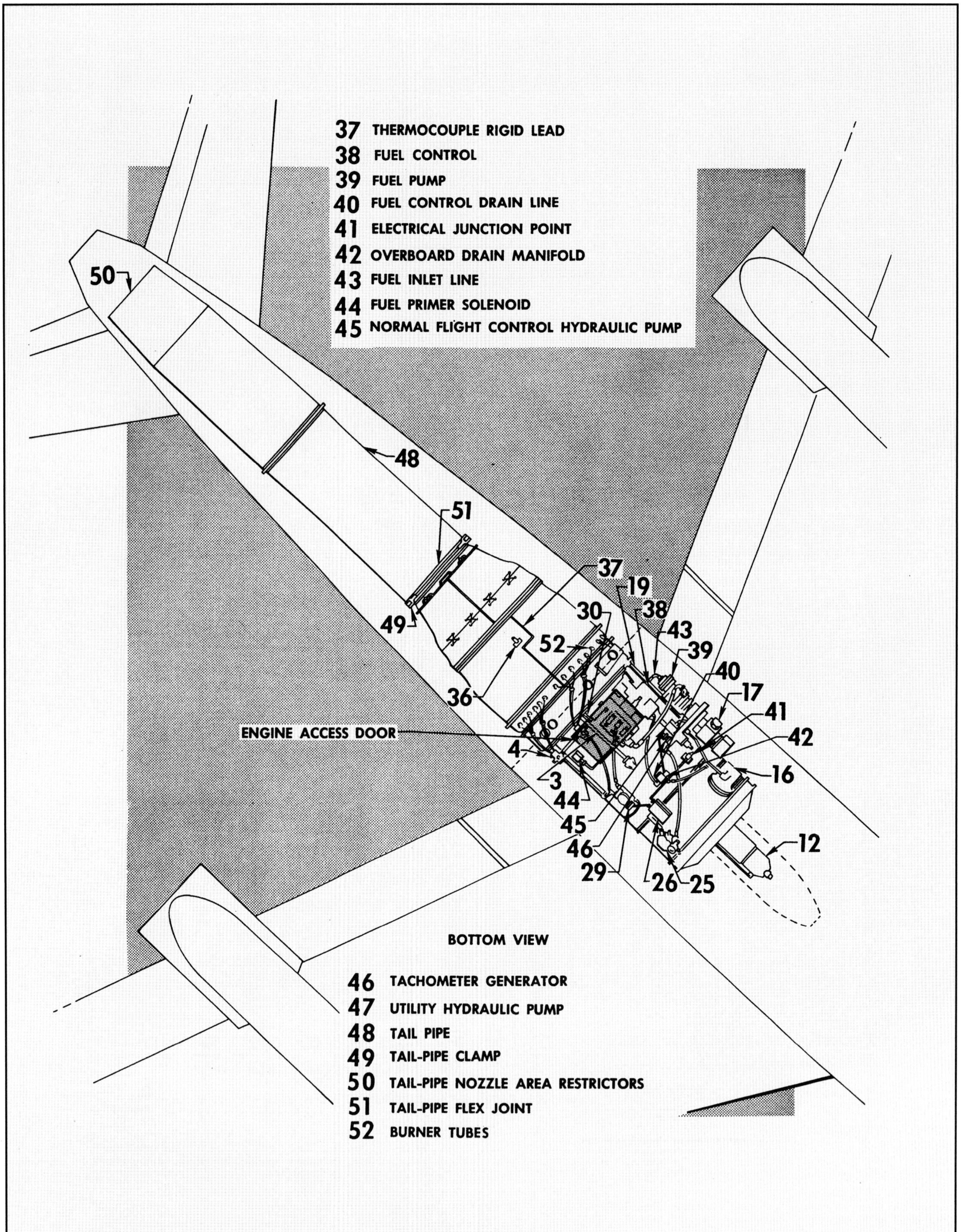

AFT FUSELAGE REMOVAL AND HANDLING WITH ADJUSTABLE DOLLY LOCKHEED 408600

To support the aft fuselage, use the ground handling dolly with adapters. Although suspending the aft fuselage by sling affords greater ease and mobility during aft fuselage removal and installation procedures, the use of the ground handling dolly with adapters is recommended for shipboard practices because it is less affected by the roll and pitch of the ship. The forward attachment fittings are installed on the airplane by removing three screws, indicated by markings on the aft fuselage, just forward of the lower arm of the speed brake, and by bolting the fittings securely at these three points. The aft attach stud is inserted in the mooring ring boss at the extreme aft end of the fuselage.

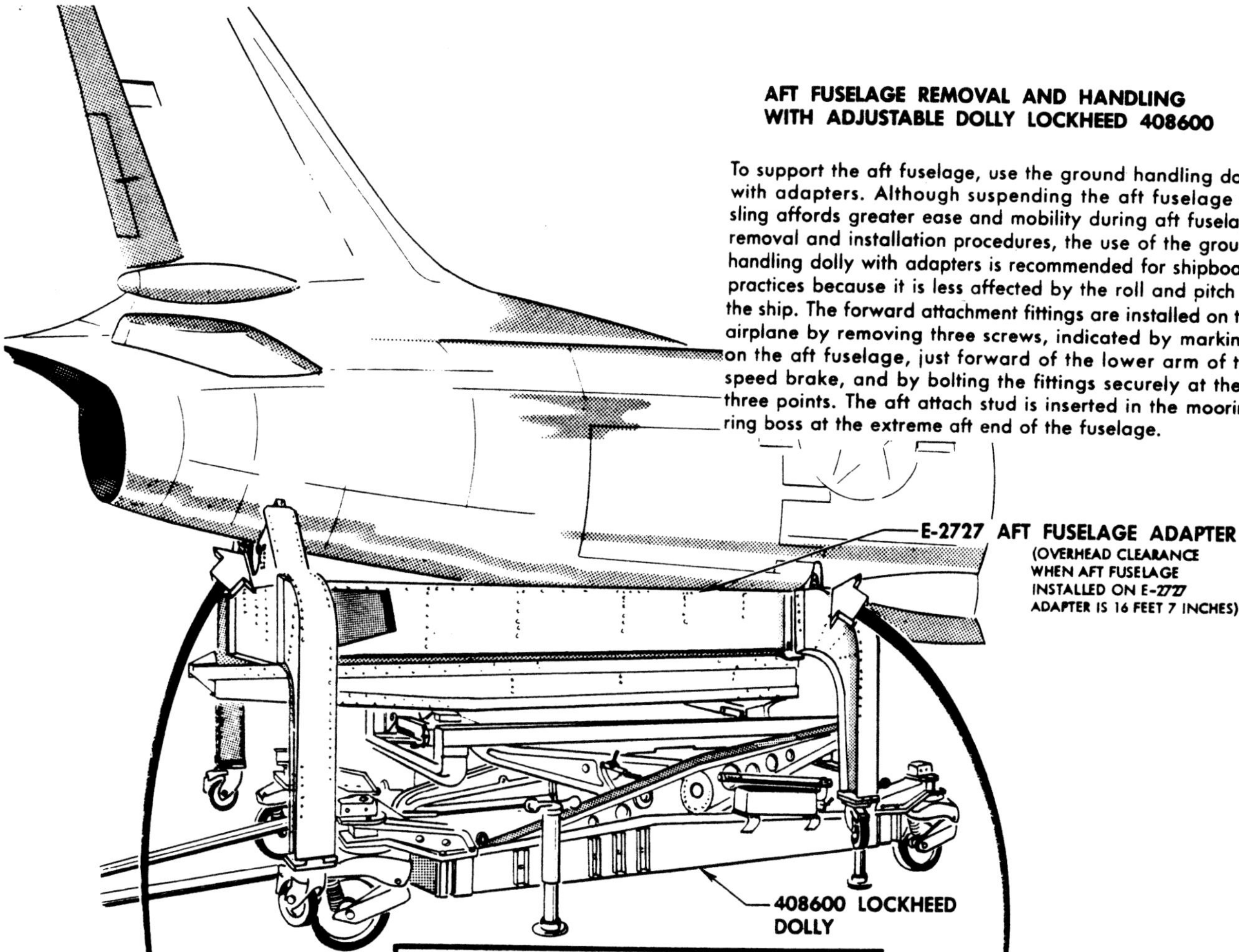

CAUTION: To use E-2727 aft fuselage adapter, airplane must be level. Completely deflate nose gear strut and fully inflate main gear struts or jack airplane to level position.

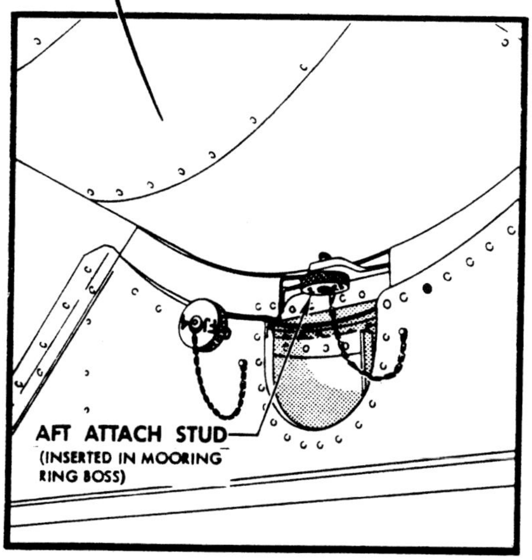

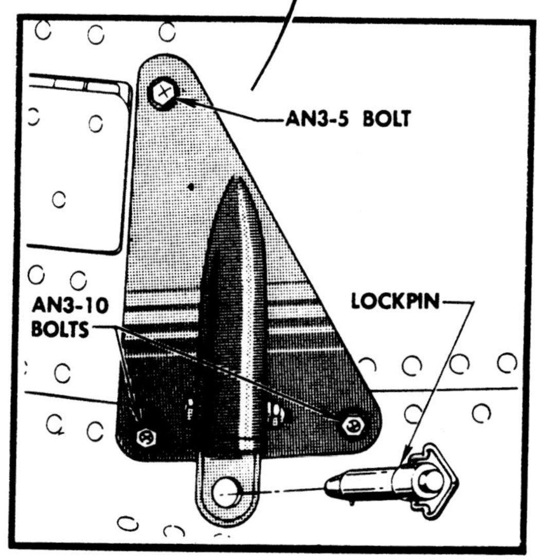

REAR FUSELAGE CRADLE ATTACHMENT FITTING

PRESSURE REFUELING WET LEADING EDGE WITH TYPE I AUXILIARY TANKS [1]

- When single-point refueling with the wings folded, the weight of the fuel in the wet leading edge would allow fuel to be forced through the wet leading edge air pressure line to the auxiliary tank and in time, completely empty the outer panels.
- To fly the airplane with auxiliary tanks empty, the wings must be extended prior to single-point refueling and not folded before flight.
- To completely fill the fuel system the auxiliary tanks should be filled first and then the internal system.

CAUTION: To prevent damage to the flexible wing fold bellows it is necessary that the wet leading edges be depressurized prior to folding the wings. Never pressurize while the wings are folded.

TYPE II AUXILIARY TANKS [1]	FILLING RATE	FILLING TIME
Internal tanks only	140 gpm	3.2 minutes (435 gallons)
Internal and wet leading edge	124 gpm	4.5 minutes
Internal, wet leading edge and auxiliary tanks	120 gpm	8.0 minutes
Wet leading edge	80 gpm	1.5 minutes
Auxiliary tanks	90 gpm	4.5 minutes
Wet leading edge and auxiliary tanks	95 gpm	5.5 minutes

NOTE: Maximum permissible fuel pressures are 70 psi from 0- to 170-gpm flow and 50 psi at flow rates greater than 170 gpm.

[1] FOR FILLING TYPE 1 AND TYPE 2 AUXILIARY TANKS (REFER TO PARAGRAPH 1-67)

FUEL SYSTEM 6-3 WET WING

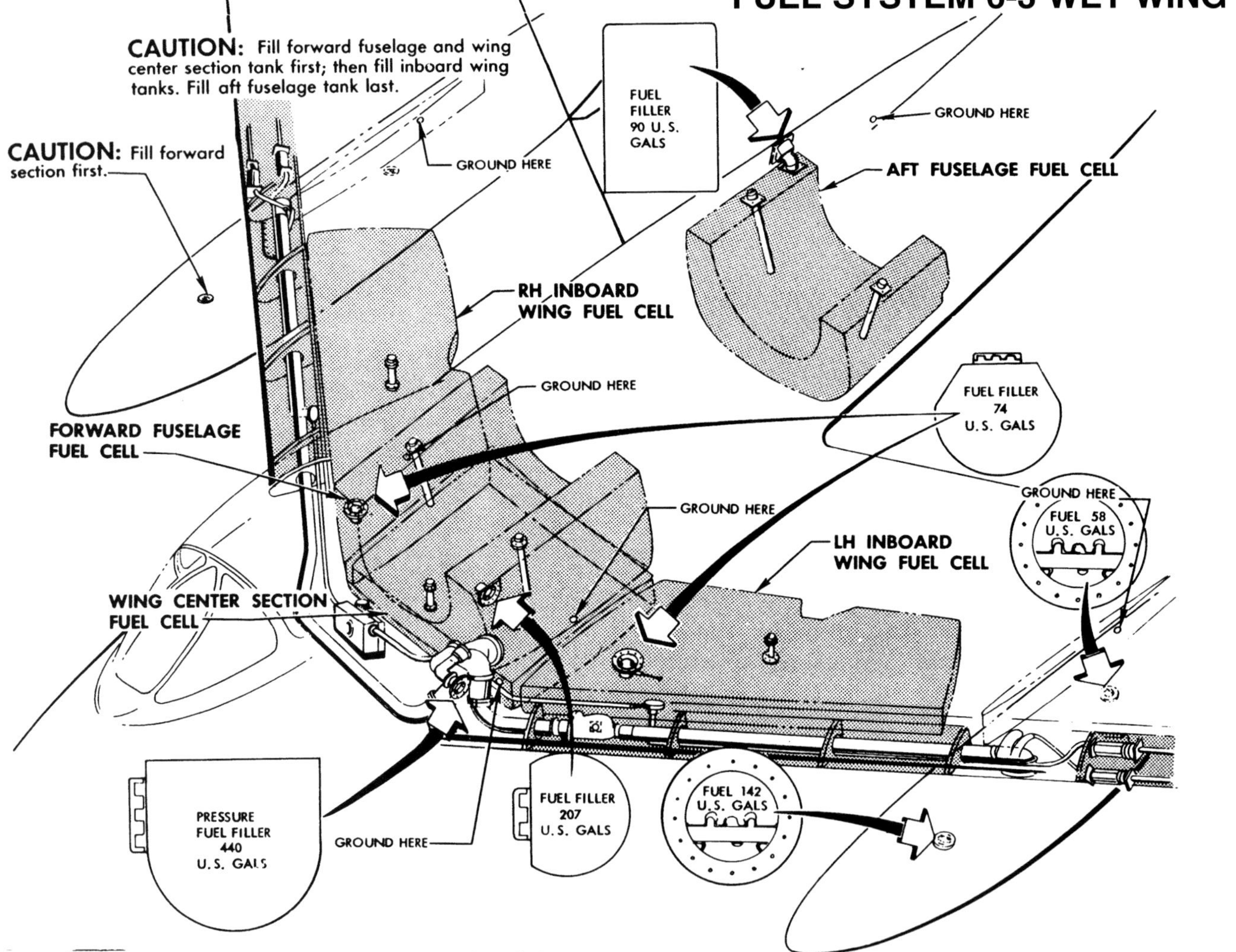

NOTE: The Type II (P/N 181-484149) auxiliary fuel tank cannot be serviced through the single-point receptacle and has no provisions for in-flight refueling. This tank is serviced through a single filler on the top portion of the tank.

150 gal. drop tanks

NOTE: The Type I auxiliary fuel tank cannot be serviced through the single-point receptacle and has no provisions for in-flight refueling. This tank is serviced through two filler caps on top portion of tank.

200 gal. drop tanks

FJ-3 WING FOLD MECHANISM

The switch to operate the wing fold was located on the right side of the cockpit below the windshield bow. A manual pin locking handle located over the switch served as a switch guard. Pulling this handle out released the wing fold pin locks and permitted access to the wing fold switch. The two-position wing fold switch was marked spread and fold. The wing fold selector valve was located in the right forward fuselage and was accessible through a door marked "Hydraulic Equipment". A lock valve, located adjacent to the selector valve, blocked return flow through the wing fold line when the wings were folded and the selector valve was de-energized. A button on the lock valve could be depressed to allow the ground crew to manually spread the wings. An indicating flap inboard of the fold joint at each leading edge folds flush to the wing surface when pins were in place and mechanically locked.

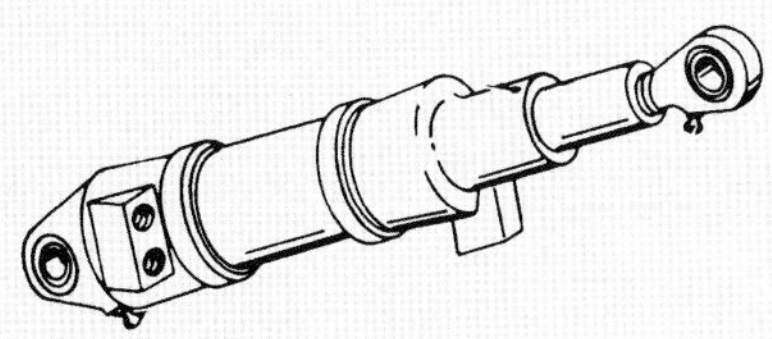

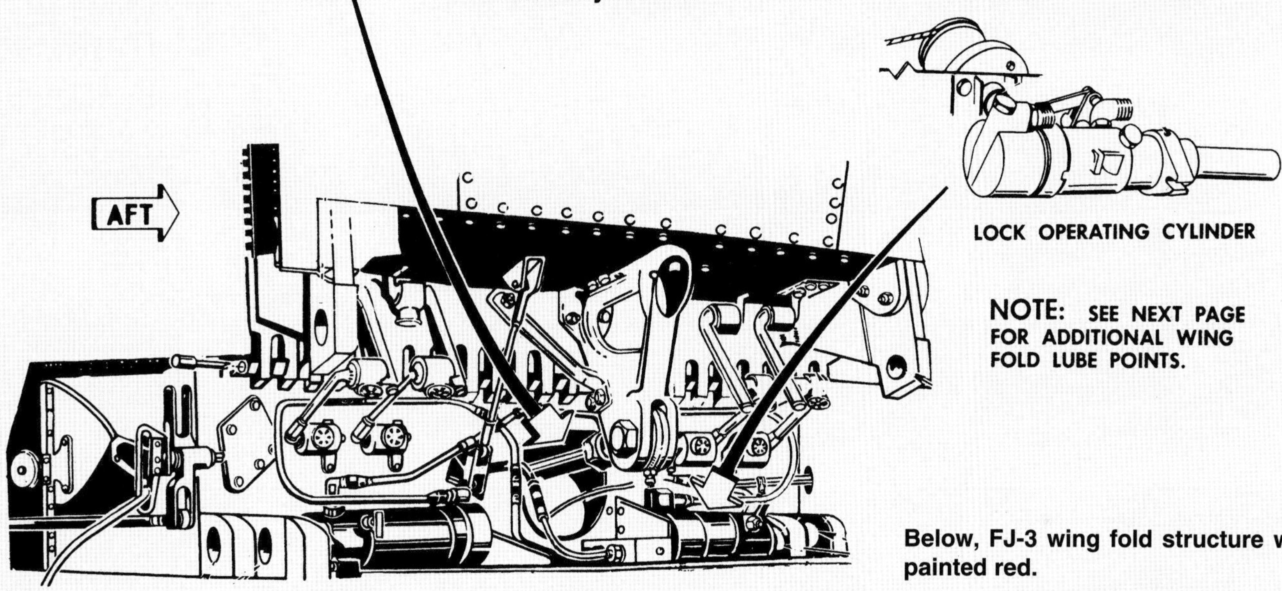

NOTE: SEE NEXT PAGE FOR ADDITIONAL WING FOLD LUBE POINTS.

Below, FJ-3 wing fold structure was painted red.

SERVICING DIAGRAM

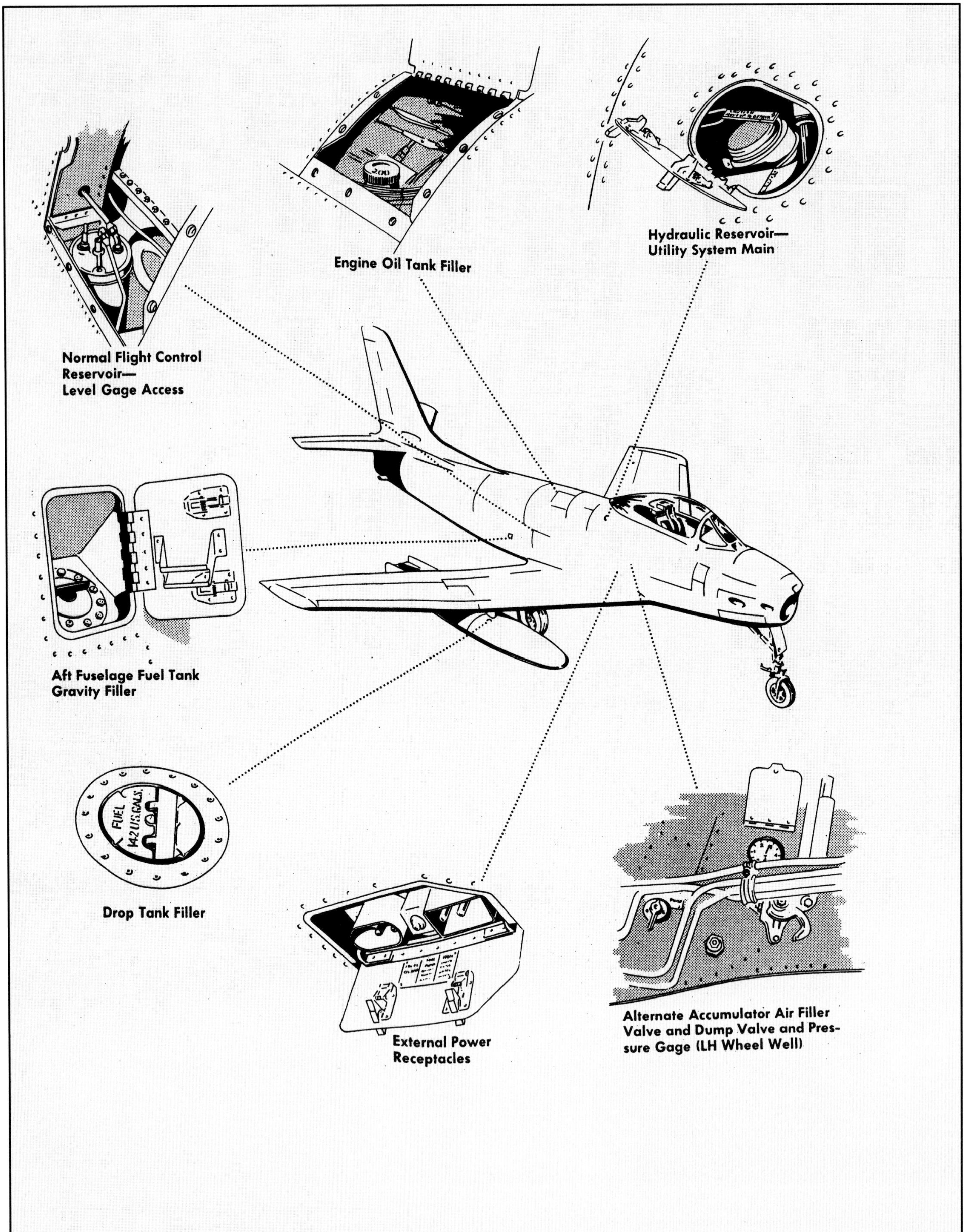

SERVICING DIAGRAM

SPECIFICATIONS

Fuel		
	Recommended	MIL-J-5624, Grade JP-5 MIL-J-5624, Grade JP-4 MIL-J-5624, Grade JP-3
	Emergency	MIL-F-5572 AvGas
Oil — Engine Lubrication	Normal Service	MIL-L-7808
	Corrosion-preventive	MIL-C-8188
Hydraulic Fluid		MIL-O-5606 (Red)

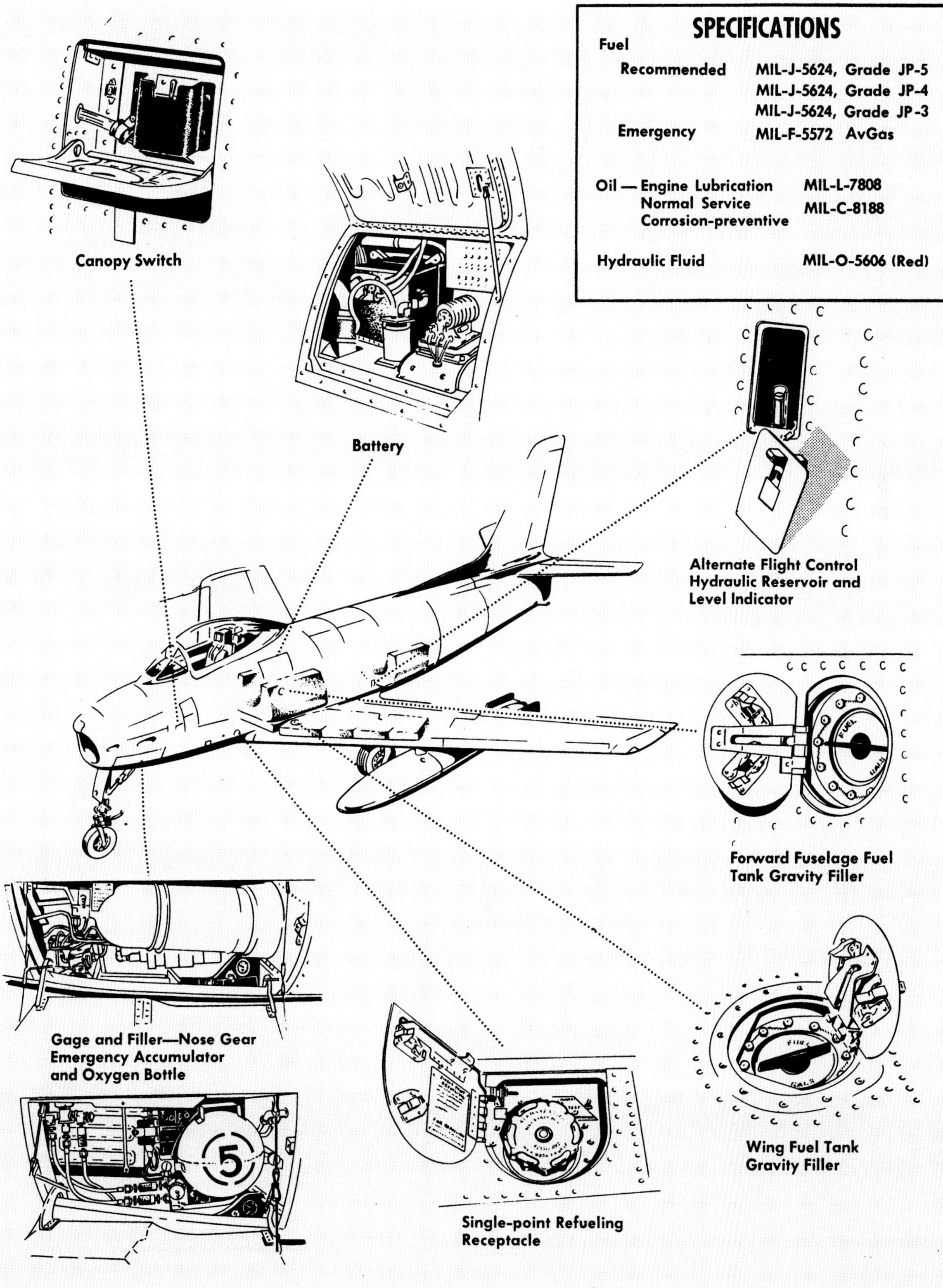

SERVICING DIAGRAM

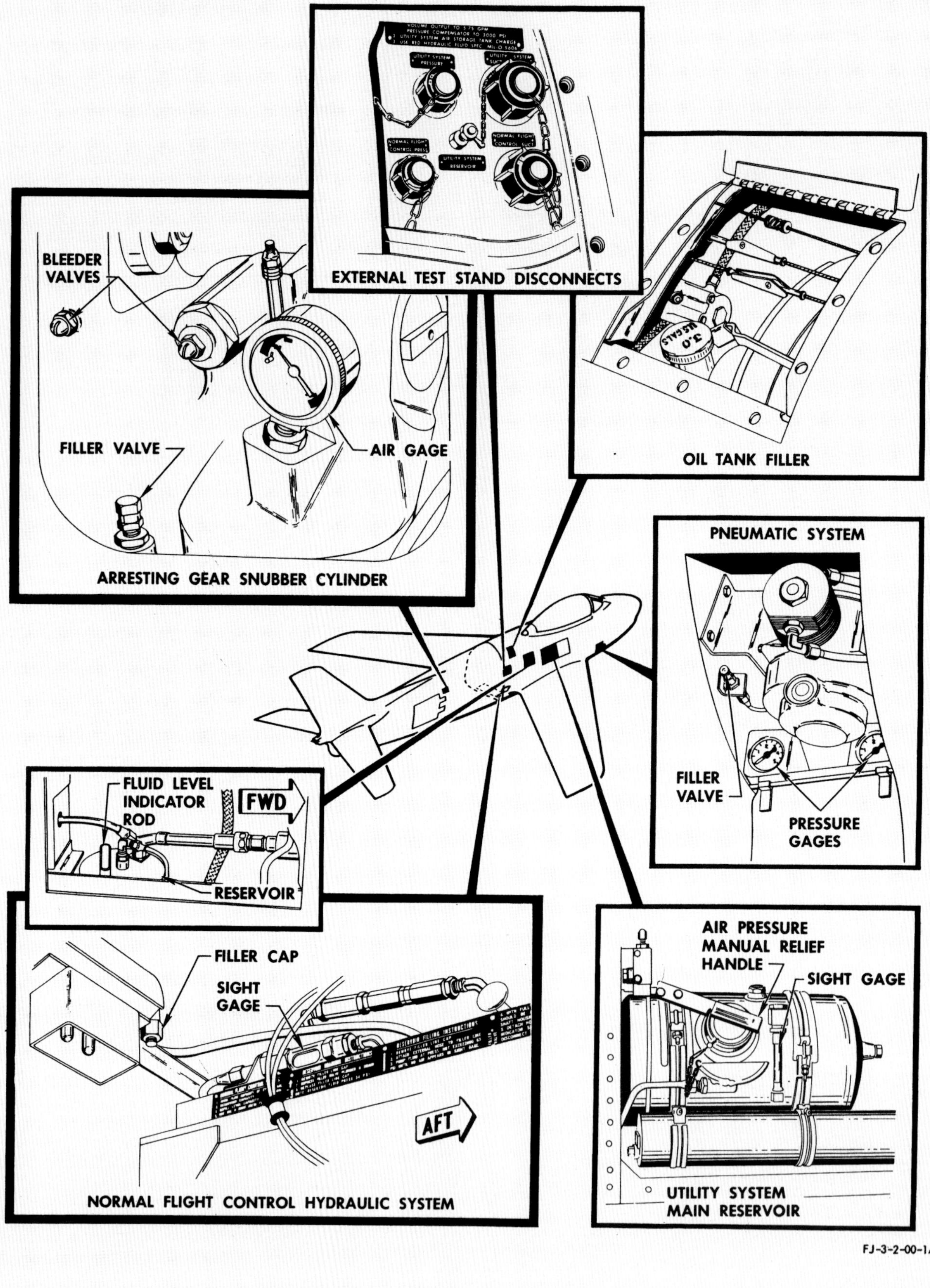

DRAINING PROVISIONS

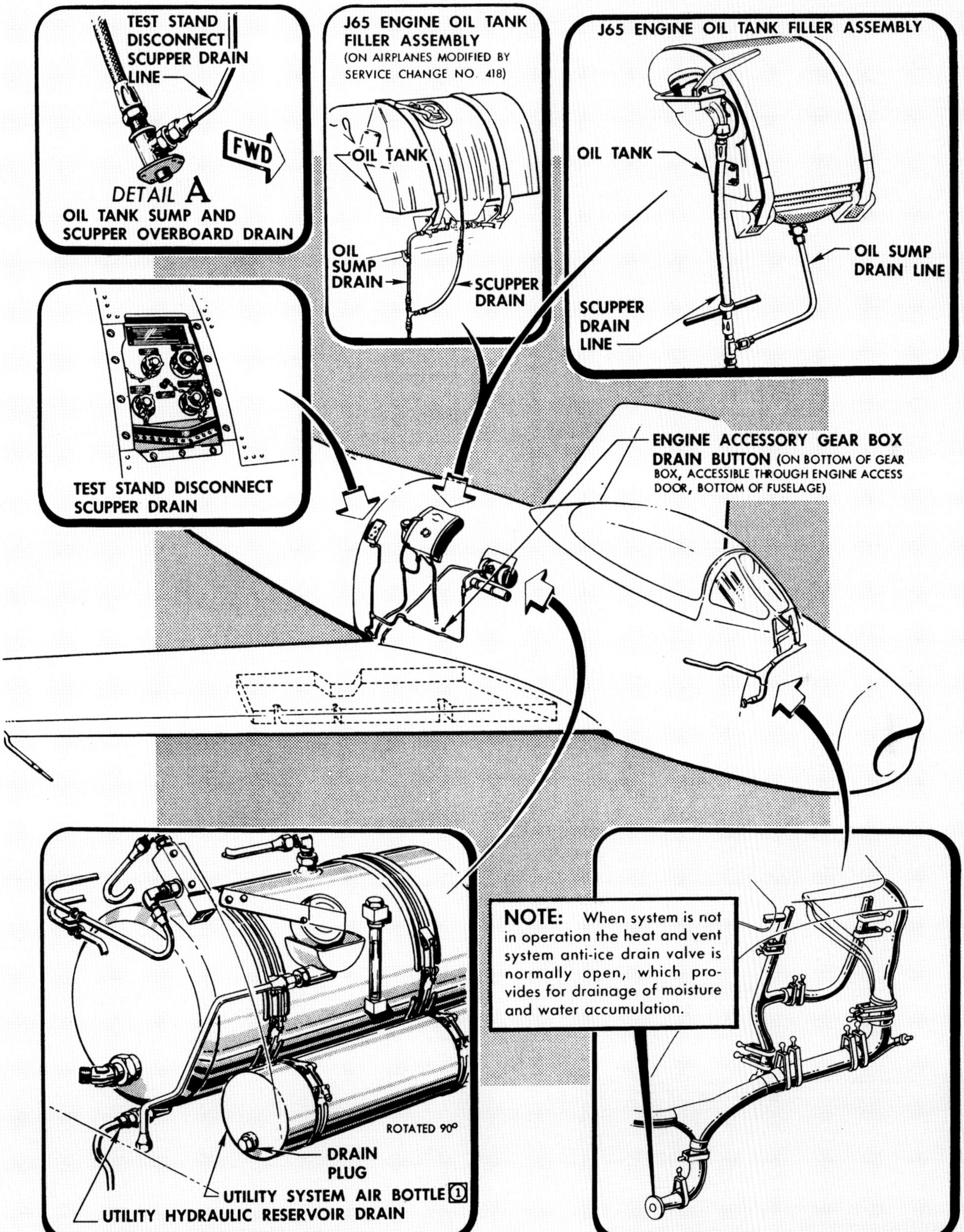

1 FJ-3a THROUGH FJ-3g AIRPLANES

SERVICING DIAGRAM

Aft end of battery compartment with refrigeration compartment aft of that.

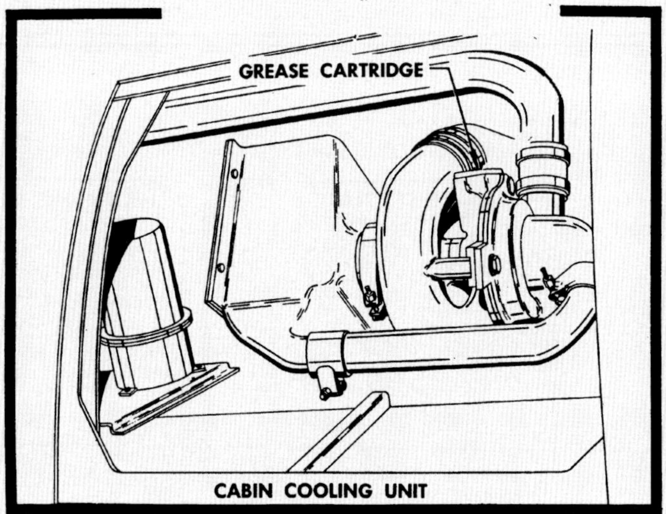

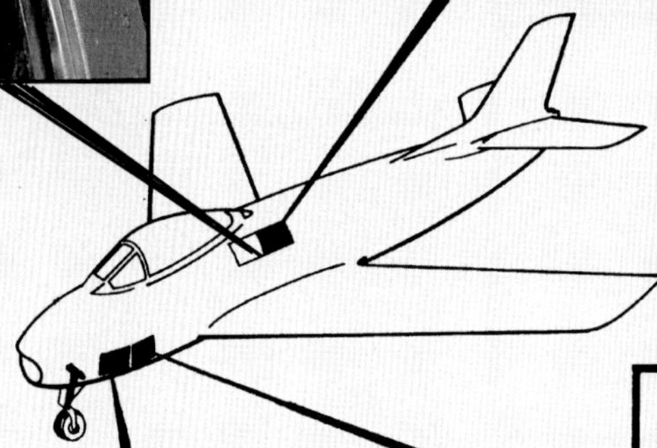

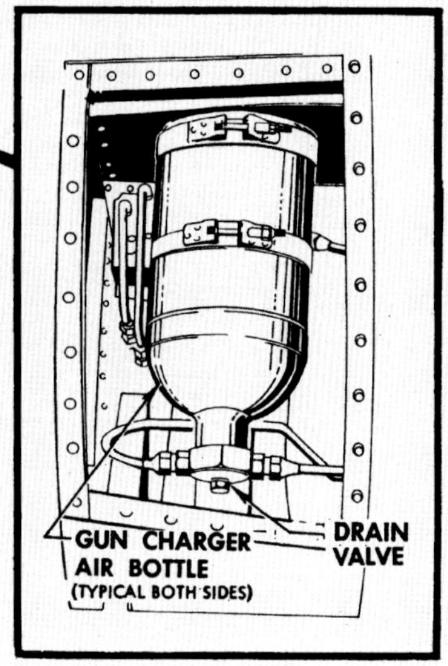

EARLY VERSION

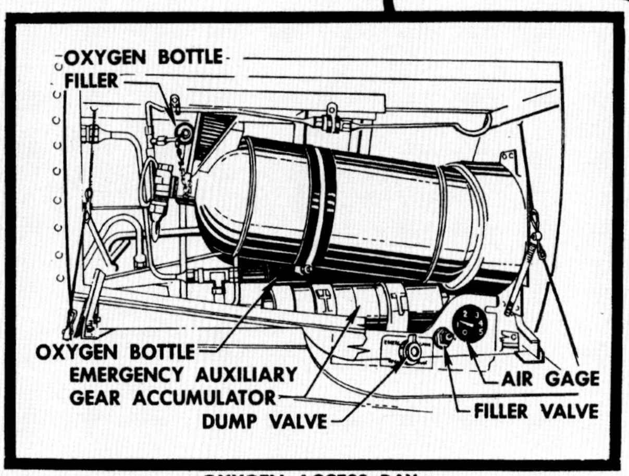

OXYGEN ACCESS BAY

LATE VERSION

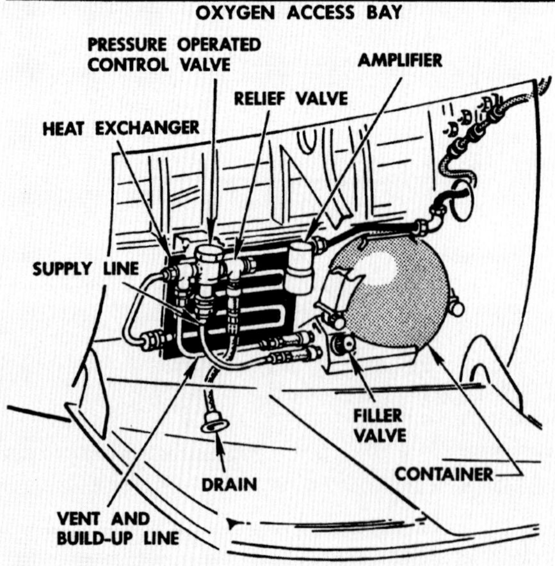

EJECTION SEAT LAYOUT

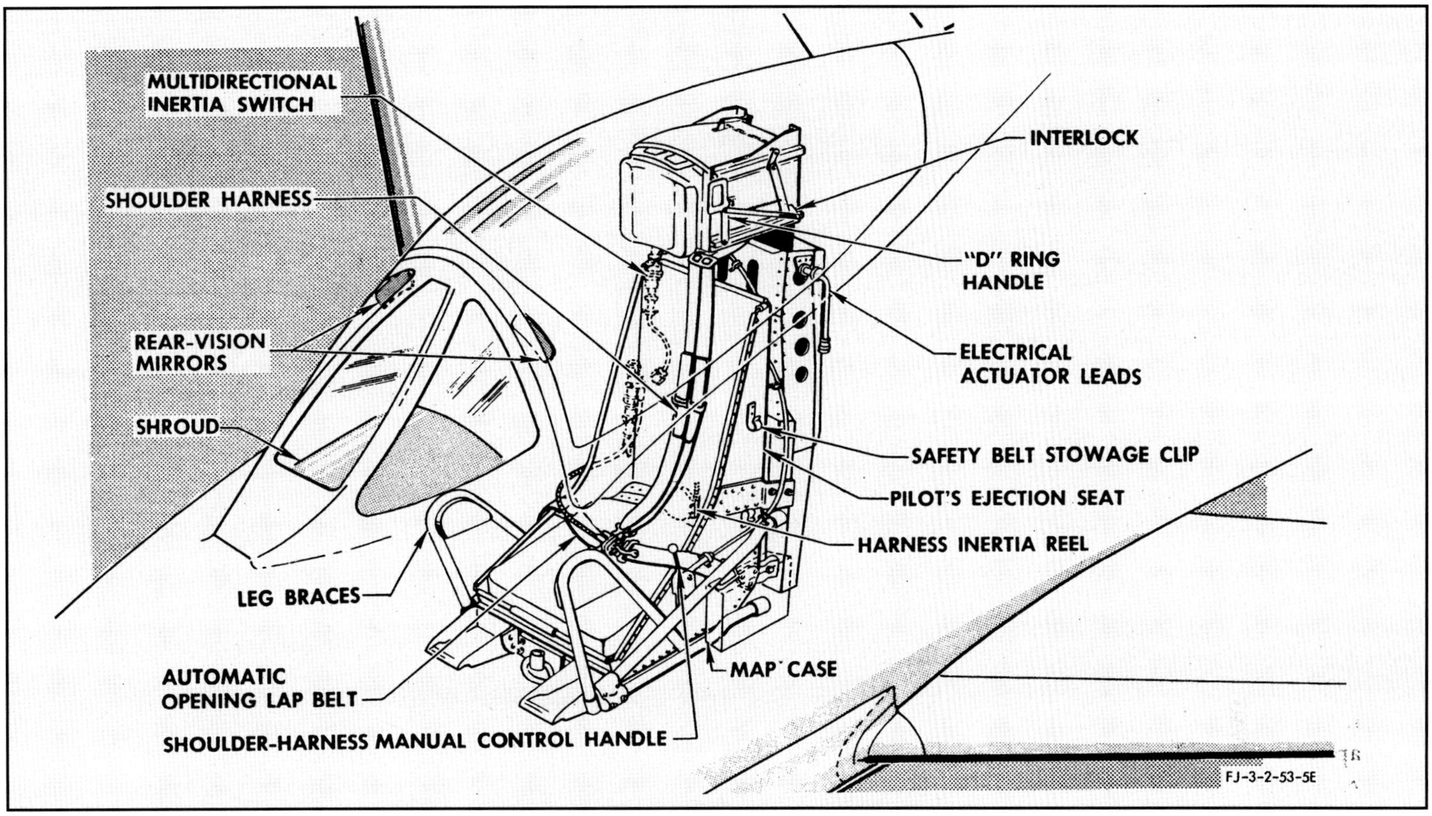

CANOPY STRUCTURES

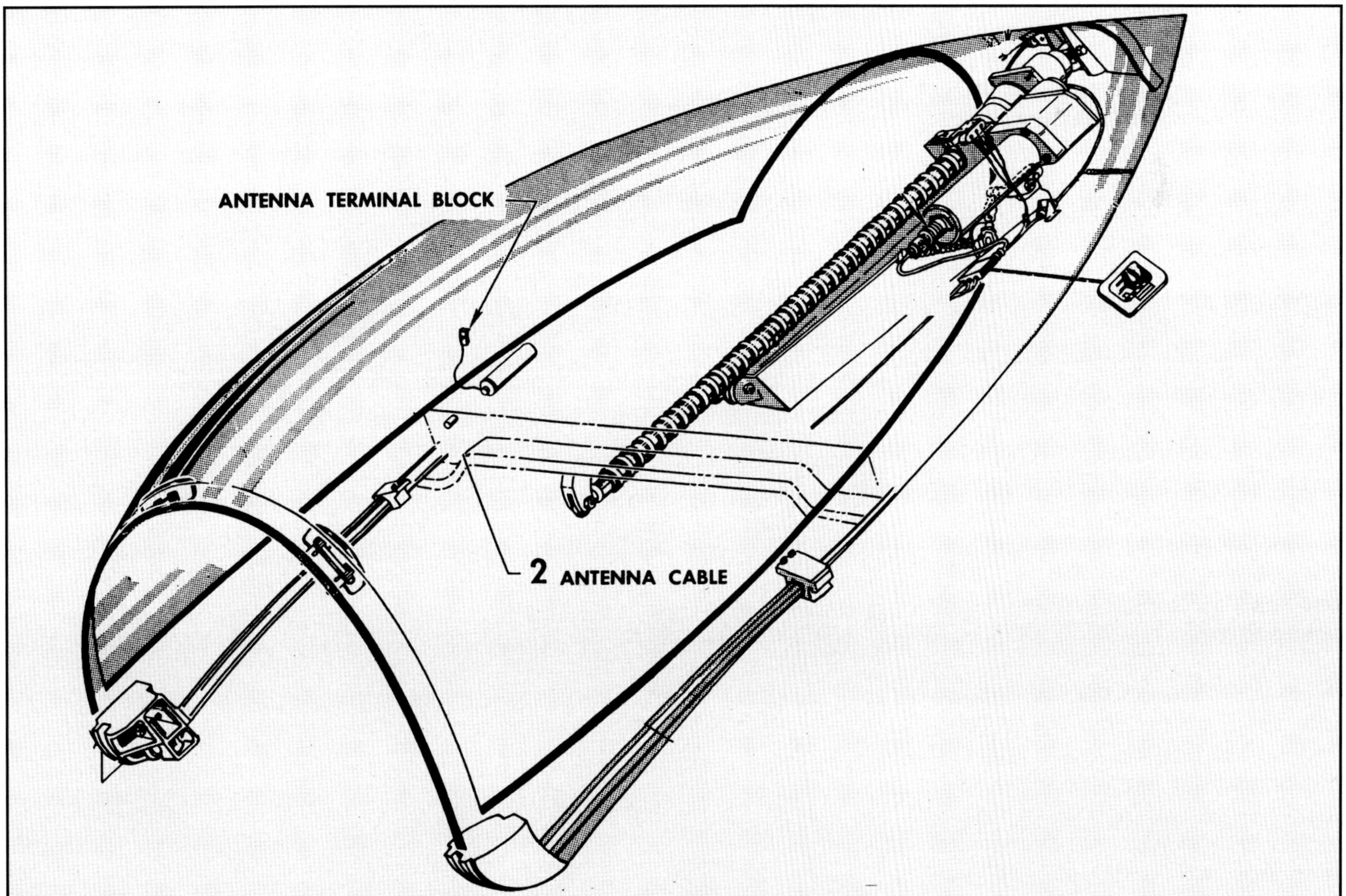

I TEST-FLEW THE FJ-3 FURY

Written for *Naval Aviation News* in 1955 by Bill Ingram, North American Aviation Test Pilot.

"The Furies, the FJ-2 and the FJ-3, have one basic system design differing from most of the other airplanes you may have flown, in that the surface controls, with the exception of the rudder which is conventional, are powered by hydraulic pressure only. The power control system is not to be confused with an hydraulic boost system. In a power control system, the pilot supplies none of the effort required to move the control surface, but merely positions a hydraulic valve at the surface to be moved, with the control stick. Since none of the loads imposed upon the control surface are transmitted back to the control stick, "feel" is supplied in flight by bungees and a bobweight.

There is a basic change in the longitudinal control surface too. The Fury has what is popularly known as a flying tail. The stick, when moved, positions the horizontal stabilizer and the elevator is geared to the stabilizer to move a predetermined amount.

The flying tail is necessary because there is a loss of control effectiveness in the transonic region, and larger control surfaces and deflections are needed.

The power control system is needed because, as airplanes become faster, more effort is required to move a control surface. This effort has become so great that it is very difficult for a conventional boost system to supply the force necessary. In fact, the power control system is rapidly becoming the normal system for high speed aircraft.

If you have not previously flown with power control, you will probably find the first time you fly the Fury that the longitudinal control is a little more sensitive than what you are accustomed to. However, this over-control tendency will disappear as you become familiar with the aircraft.

FJ-3 INSTRUMENT PANEL

Above, factory photo of an FJ-3 on 9 October 1953. Below, on 30 December 1954.

INSTRUMENT PANEL FJ-3a through FJ-3h
Not having service change number 151, 413, or 416 complied with

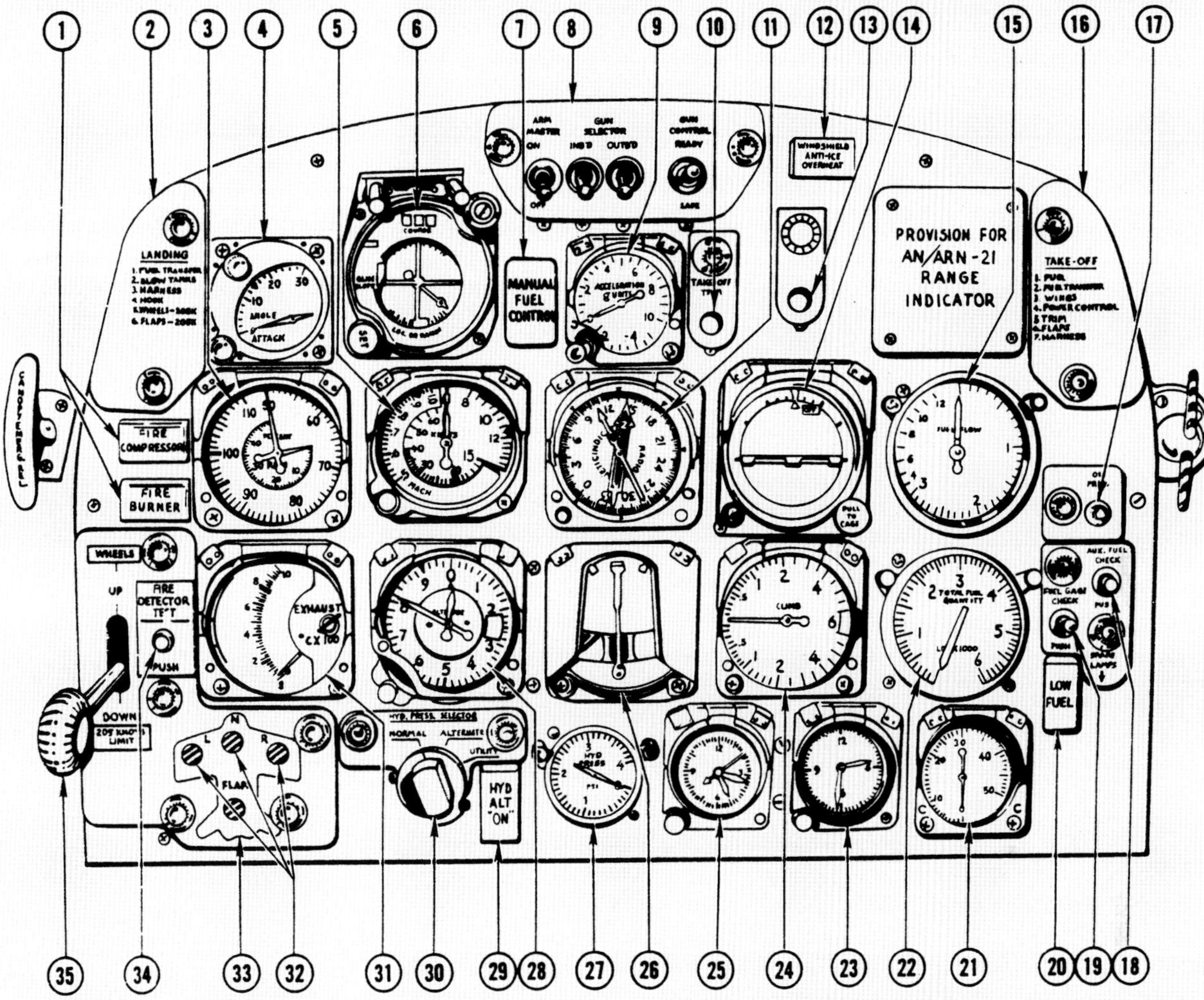

1 ENGINE FIRE DETECTOR WARNING LIGHTS
2 LANDING CHECK LIST
3 TACHOMETER INDICATOR
4 ANGLE-OF-ATTACK INDICATOR
5 AIRSPEED-MACH NUMBER INDICATOR
6 COURSE INDICATOR
7 MANUAL FUEL CONTROL WARNING LIGHT
8 GUN CONTROL PANEL
9 ACCELEROMETER
10 TAKE-OFF TRIM INDICATOR
11 RADIO MAGNETIC COURSE INDICATOR
12 WINDSHIELD ANTI-ICE OVERHEAT LIGHT
13 FUEL TRANSFER INDICATOR
14 GYRO HORIZON INDICATOR
15 FUEL FLOWMETER
16 TAKE-OFF CHECK LIST
17 OIL PRESSURE INDICATOR
18 AUXILIARY FUEL CHECK SWITCH
19 FUEL GAGE CHECK SWITCH
20 LOW FUEL LEVEL WARNING LIGHT
21 CABIN PRESSURE ALTITUDE INDICATOR
22 FUEL QUANTITY INDICATOR
23 CLOCK
24 RATE-OF-CLIMB INDICATOR
25 ELAPSED-TIME CLOCK
26 TURN-AND-BANK INDICATOR
27 HYDRAULIC PRESSURE INDICATOR
28 ALTIMETER
29 HYDRAULIC ALTERNATE SYSTEM "ON" LIGHT
30 HYDRAULIC PRESSURE SELECTOR SWITCH
31 EXHAUST TEMPERATURE INDICATOR
32 LANDING GEAR POSITION INDICATORS
33 FLAP POSITION INDICATOR
34 FIRE DETECTOR TEST SWITCH
35 LANDING GEAR CONTROL HANDLE

Above, FJ-3 on 17 November 1954. Below, FJ-3 on 1 March 1956. (USN)

Normal take-off is made with full flap and with the trim indicator showing IN for all surface controls. The aircraft flies away nicely, and take-off visibility is good. During cross-wind take-off, the up-wind wing tends to rise rather more than on a straight wing aircraft; however, this is not a problem. The flaps may be retracted soon after take-off with no settling and very little trim change.

The Fury has excellent stall characteristics both accelerated and unaccelerated. During wings-level unaccelerated stall, there is ample warning in the form of general airframe buffeting, increasing the intensity up to the point of stall. The stick can be brought back to the aft stop, if handled carefully, and the aircraft will porpoise gently. Recovery is easily made by easing forward on the stick.

The accelerated stall is preceded by ample warning in the form of buffeting. At high altitude, the buffeting will become objectionable before actual stall is reached. Recovery is easily made by relaxing stick pressure.

The airplane has excellent handling qualities at high speeds. In fact, you will find you are able to climb at speeds that are considered limit or above-limit in some other jet fighters. Stability and control are good, but here is a word of caution. At high speeds and low altitudes, the power control system is very effective, and the structural strength of the airplane can be easily exceeded by rough handling of the controls. In general, when operating at low altitudes and high speed, applications of "G" should be made smoothly.

I have already mentioned the loss of longitudinal control effectiveness in the transonic region. This phenomenon is true of the ailerons also. A reduction of aileron effectiveness is manifested by the phenomenon popularly called wing roll. This is an abrupt lateral trim change which occurs just under sonic speed and is normal. However, acceleration and satisfactory control can be regained after passing through the "wing roll".

Remember that in high speed dives, "wing roll" will be encountered twice: once when accelerating through and once when decelerating back during pull-out. Wing roll is a function of Mach and will vary with altitude. At high altitudes, it is less severe and occurs at slightly higher Machs.

The speed brakes may be extended at any airspeed. Extending speed brakes causes a mild pitch-up, but it is easily controlled, and you will probably want to retrim the aircraft. As you would expect, retracting speed brakes causes slight nose heaviness.

The Furies have excellent spinning characteristics. It is necessary to force the aircraft to spin. The wings level entry produces an oscillatory, poorly developed spin while an entry from a turn will give a more "normal" spin. Recovery is easily accomplished by neutralizing the controls in both cases.

The Fury is an exceptionally easy airplane to land, and its low speed handling characteristics are excellent. Good control is maintained, about all axes, down to actual stall. The forward visibility is superior, to most fighters flown by this writer, either for carrier landings or field landings. The position of the speed brakes is a matter of opinion; however, the writer prefers to approach with speed brakes extended.

The flight characteristics are essentially the same for both the FJ-2 and FJ-3. However, the FJ-3 has a decided edge in performance, especially climb and acceleration to maximum speed.

The major difference between the two Furies is the power plant. The FJ-2 has a GE J47 engine with an official thrust rating of about 6,000 lbs; and the FJ-3, the Wright J65 engine with a thrust rating of about 8,000 lbs.

I hope that you "fighter-types" will have as much fun flying the Fury series as we have had at North American during design & development."

Above and below, factory photos of FJ-3 BuNo 346 on 25 July 1955. (USN)

INSTRUMENT PANEL FJ-3i and subsequent and all airplanes with service change number 151, 413, 416, 433 and 460 complied with

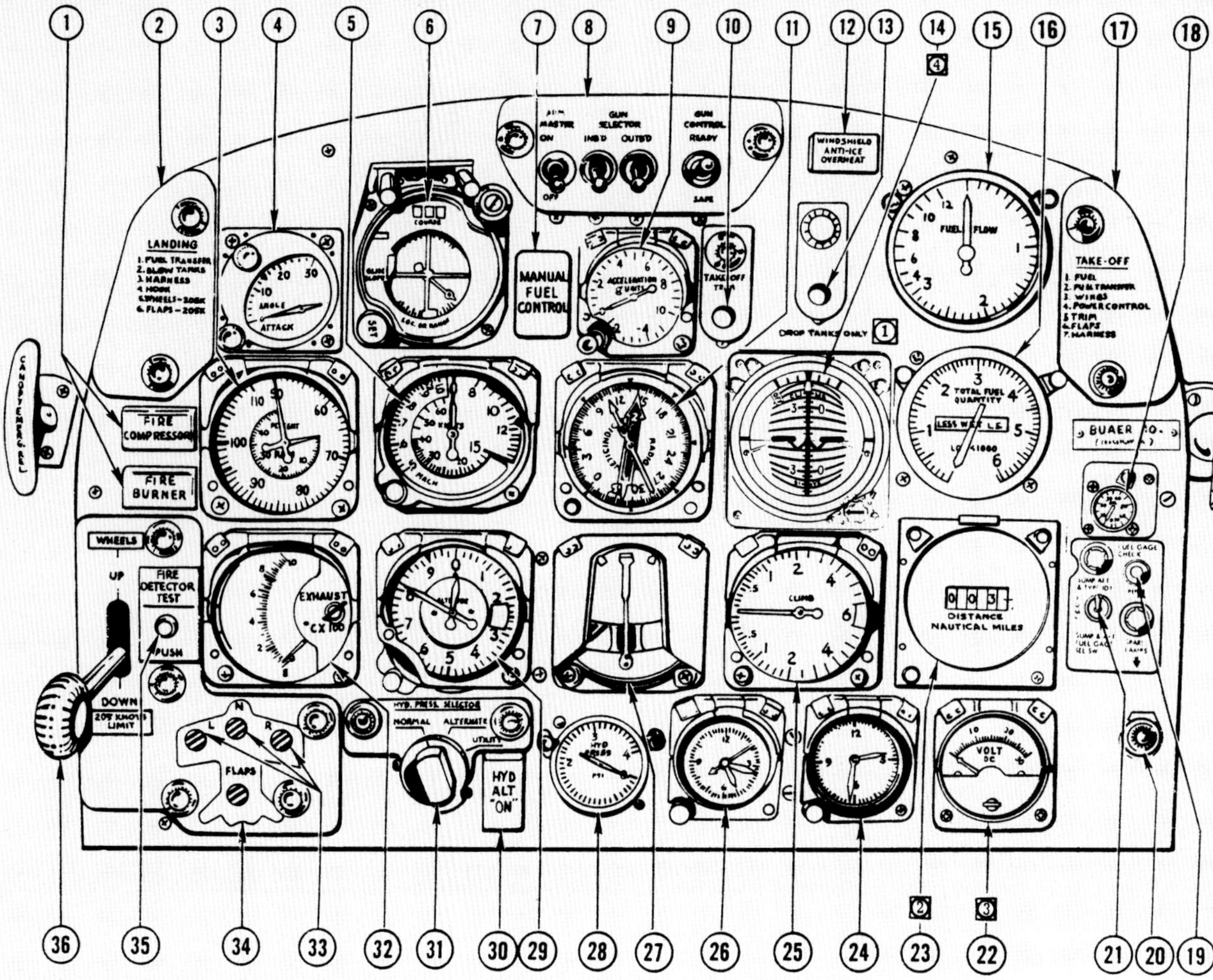

1 ENGINE FIRE DETECTOR WARNING LIGHTS
2 LANDING CHECK LIST
3 TACHOMETER INDICATOR
4 ANGLE-OF-ATTACK INDICATOR
5 AIRSPEED-MACH NUMBER INDICATOR
6 COURSE INDICATOR
7 MANUAL FUEL CONTROL WARNING LIGHT
8 GUN CONTROL PANEL
9 ACCELEROMETER
10 TAKE-OFF TRIM INDICATOR
11 RADIO MAGNETIC COURSE INDICATOR
12 WINDSHIELD ANTI-ICE OVERHEAT LIGHT
13 FUEL TRANSFER INDICATOR
14 REMOTE ATTITUDE INDICATOR [4]
15 FUEL FLOWMETER
16 FUEL QUANTITY INDICATOR
17 TAKE-OFF CHECK LIST
18 OIL PRESSURE INDICATOR
19 FUEL GAGE CHECK SWITCH
20 SPARE LAMP HOLDER
21 FUEL GAGE CHECK SWITCH
22 VOLTMETER [3]
23 AN/ARN-21 RANGE INDICATOR [2]
24 CLOCK
25 RATE-OF-CLIMB INDICATOR
26 ELAPSED-TIME CLOCK
27 TURN-AND-BANK INDICATOR
28 HYDRAULIC PRESSURE INDICATOR
29 ALTIMETER
30 HYDRAULIC ALTERNATE SYSTEM "ON" LIGHT
31 HYDRAULIC PRESSURE SELECTOR SWITCH
32 EXHAUST TEMPERATURE INDICATOR
33 LANDING GEAR POSITION INDICATORS
34 FLAP POSITION INDICATOR
35 FIRE DETECTOR TEST SWITCH
36 LANDING GEAR CONTROL HANDLE

[1] AIRPLANES HAVING SERVICE CHANGE NO. 416 COMPLIED WITH
[2] AIRPLANES HAVING SERVICE CHANGE NO. 151 COMPLIED WITH
[3] AIRPLANES HAVING SERVICE CHANGE NO. 413 COMPLIED WITH
[4] AIRPLANES HAVING SERVICE CHANGE NO. 433 COMPLIED WITH

FJ-3-2-51-24C

SPINNING THE FJ-3 FURY

Written for *Approach* magazine in 1955 by Jim Pearce, North American Aviation Test Pilot.

The FJ-2 and FJ-3 Fury have been thoroughly spun with the gear and flaps up and down, speed brakes open and closed, and under each of these conditions with the 200 gallon external tanks off, on and empty, and on and full.

The airplanes have consistently demonstrated their ability to recover from any of those spins using standard recovery techniques. During all the spin testing conducted by North American Aviation Inc., on the Fury as well as its Air Force counterpart, the Sabrejet, no spin has ever been encountered from which recovery could not be promptly accomplished by normal spin recovery techniques. In fact it is difficult to hold an FJ in spins under certain conditions.

Spins can be encountered by stalling the airplane in level flight and applying hard rudder one direction or the other simultaneous with full back

FJ-3 PILOT'S LEFT-HAND CONSOLE

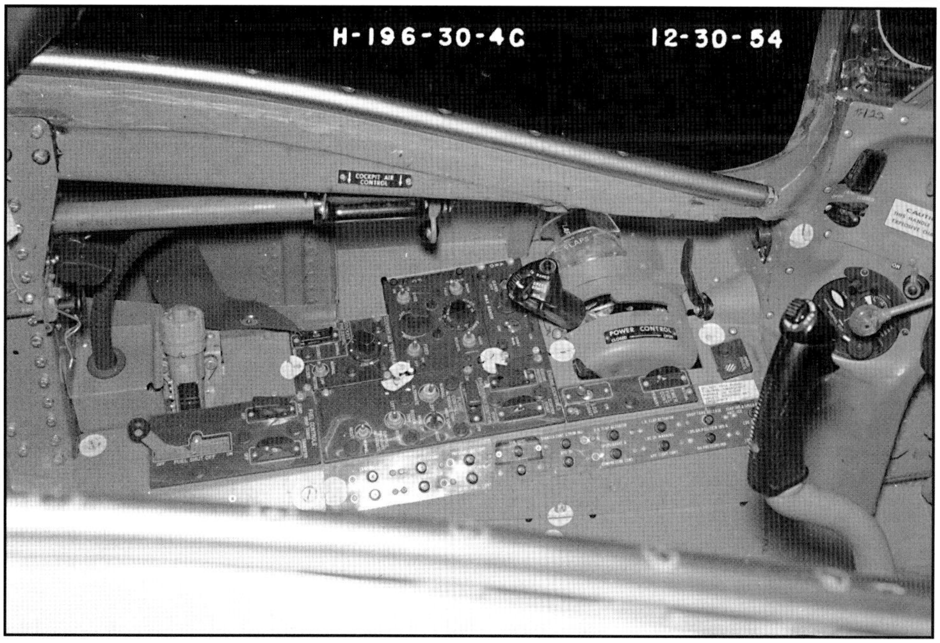

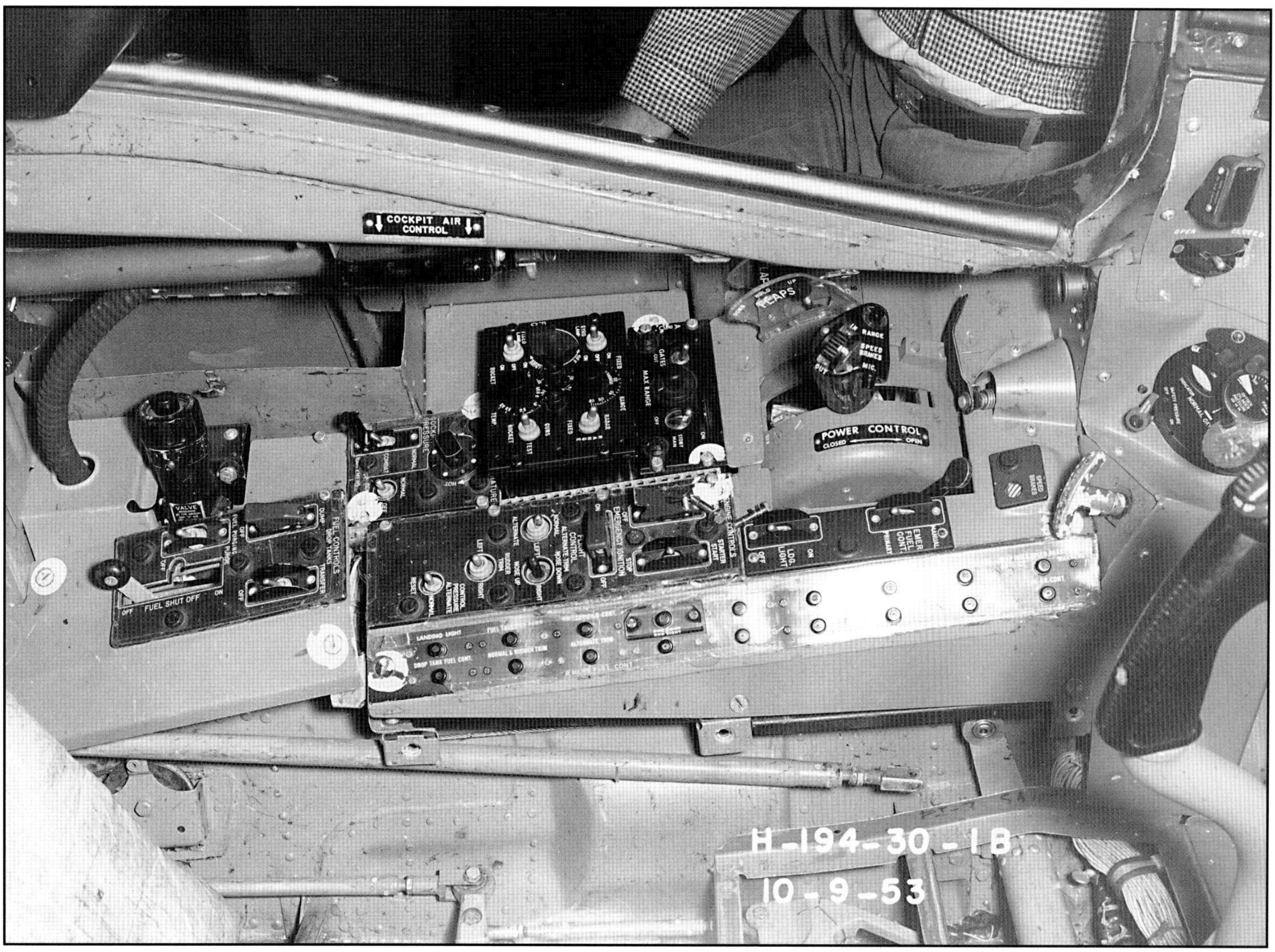

FJ-3 PILOT'S LEFT-HAND CONSOLE

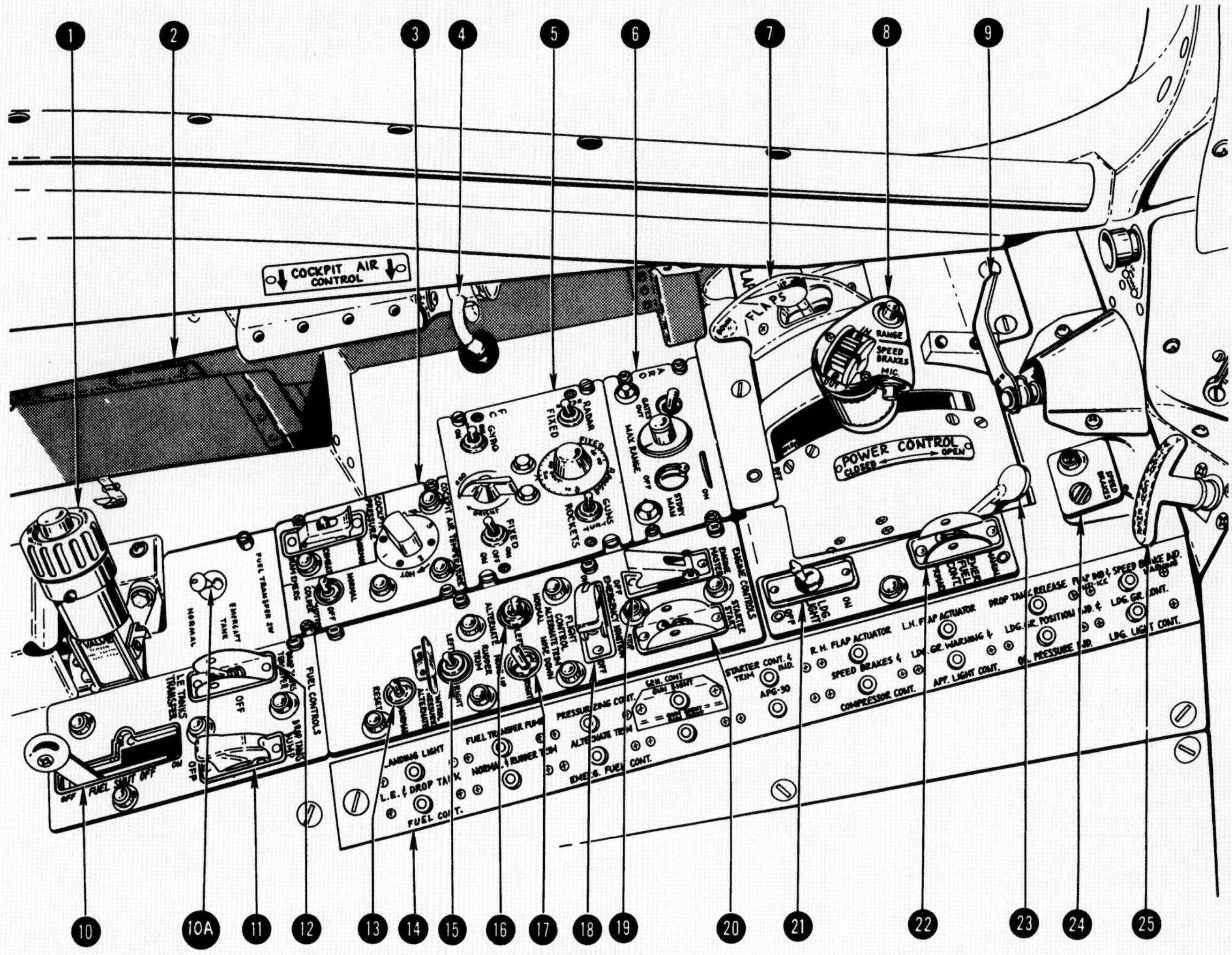

LEFT-HAND CONSOLE

1. Anti-G Suit Regulator Valve
2. Map Case
3. Cockpit Air Temperature Control Panel
4. Cabin Air Deflector Control
5. AFCS Control Panel
6. Radar Set Control Panel
7. Wing Flap Control
8. Throttle
9. Throttle Catapult Handle
10. Fuel Shutoff Control
11. Drop Tank Fuel Dump Switch
10A. Emergency Aft Tank Fuel Transfer Switch*
12. Drop Tank Transfer Switch
13. Flight Control System Selector Switch
14. Circuit-breaker Panel
15. Rudder Trim Switch
16. Trim Control Selector Switch
17. Stabilizer-aileron Alternate Trim Switch
18. Emergency Ignition Switch
19. Engine Master Switch
20. Starter Start-stop Switch
21. Landing Light Switch
22. Emergency Fuel Control Switch
23. Throttle Friction Lock
24. Speed Brake Position Indicator
25. Manual Emergency Change-over Handle

* AIRPLANES HAVING ASC 536 COMPLIED WITH

FJ-3-1-00-3C

stick. Or they can be encountered during turning flight at any speed up to .9 Mach number, at the higher altitudes, by pulling too tight in a turn, causing an accelerated stall, and simultaneously applying rudder one direction or the other. If spins should be encountered under either of these conditions, immediate neutralizing of all controls will halt the spin if this action is taken before one complete turn is completed. If a fully developed spin is allowed to occur by maintaining pre-spin controls recovery can still be affected using standard recovery techniques.

REMOVE AFT TRIM

Remember to recover from any type of spin in an FJ aircraft the pilot should neutralize the stick and ailerons immediately, and simultaneously apply hard anti-spin rudder. Neutralizing the stick does not necessarily mean returning it to its trim position for the FJ airplane will not recover from a spin unless the stabilizer is

FJ-3 PILOT'S LEFT-HAND CONSOLE

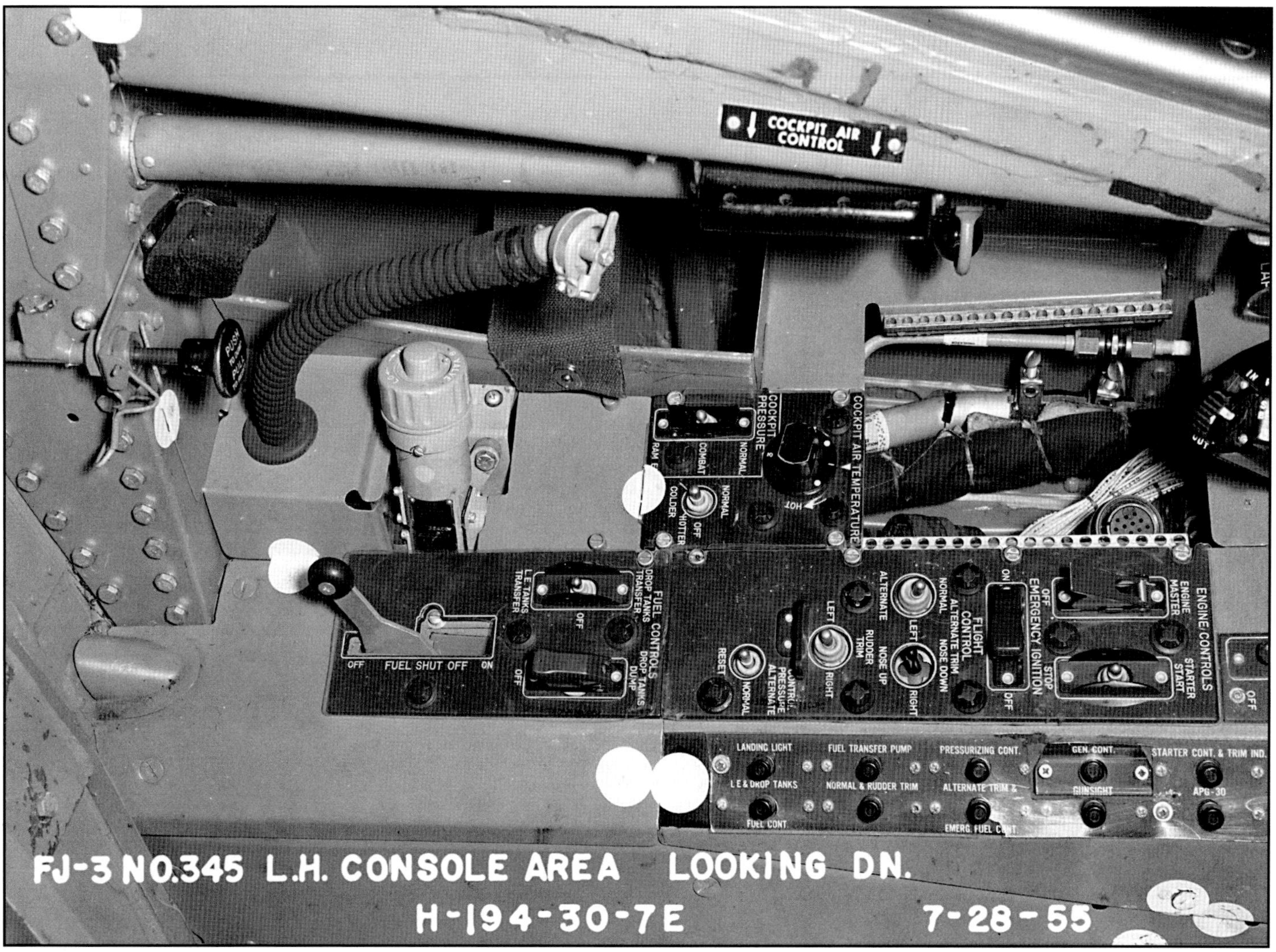

FJ-3 PILOT'S RIGHT-HAND CONSOLE

moved further forward than the stabilizer setting obtained with full aft trims; otherwise, reversing the rudder will merely reverse the direction of the spin.

A maximum force of only 40 pounds is required to move the stick fully forward when the airplane is trimmed full nose-up and even in a well developed spin full anti-spin rudder requires only 160 to 200 pounds. Therefore, there are no unmanageable forces involved in affecting satisfactory spin recoveries.

It is not essential that power be reduced after entry into a spin but the reduction of power to idle reduces the altitude loss per turn.

If, at any time a pilot in a spin becomes confused, the best thing for him to do is to release all controls, provided that he has not trimmed into a turn prior to spin entry. A good rule

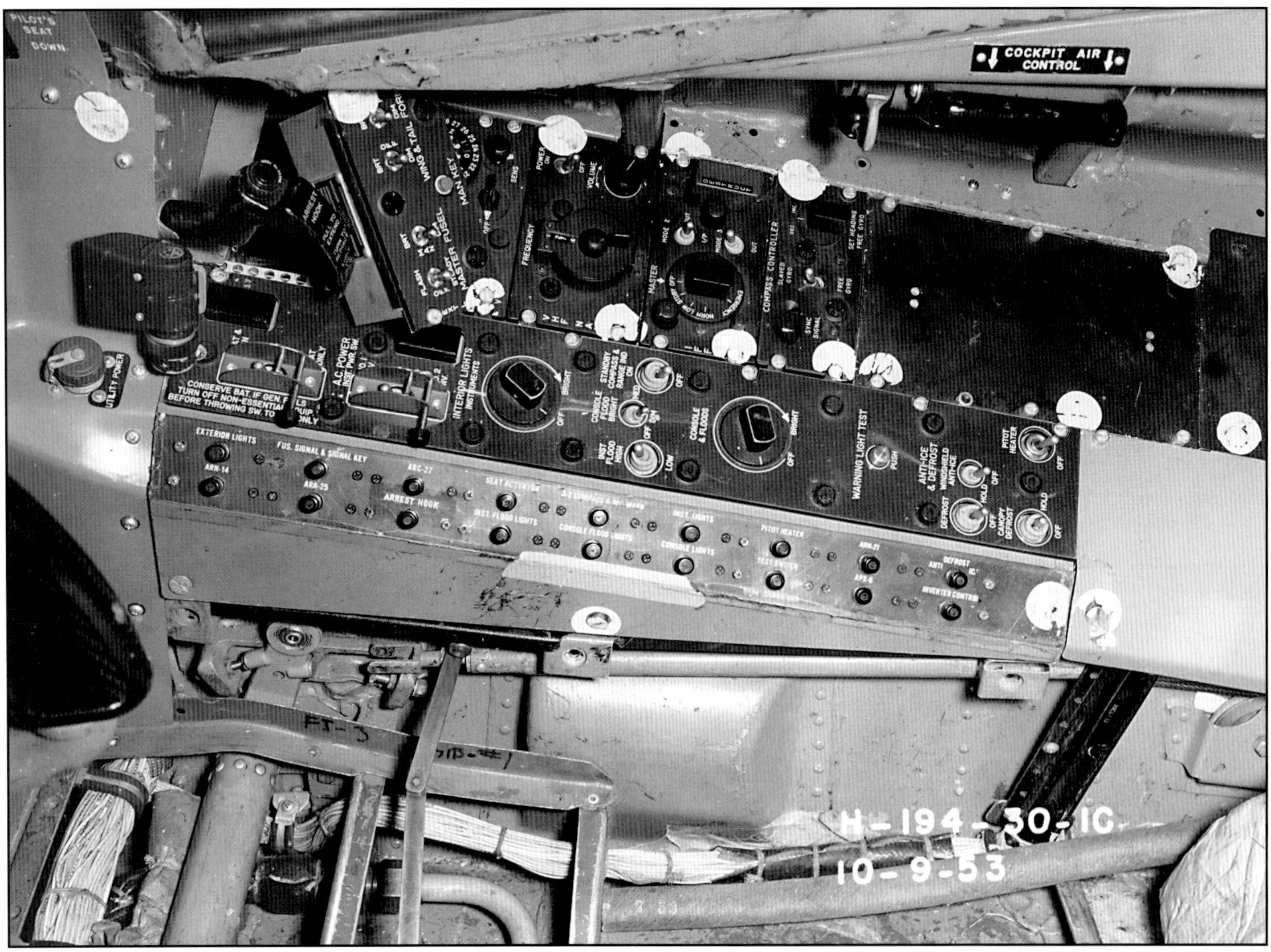

of thumb is to never find yourself in this situation by never trimming into a turn.

INVERTED SPINS

Inverted spins might occur if the airplane should stall during the inverted portion of a maneuver. The inverted spin can be recognized by a roll from an inverted position over to what appears to be a normal 45° dive during each turn. Recovery can be made at any time by neutralizing the controls and flying the airplane out of the spin at the time that it is upright.

Spins with drop tanks installed are for all practical purposes, identical to clean airplane spins except that if rudder is not purposely applied as the airplane stalls in an accelerated turn, the airplane tends to perform a snap roll in the opposite direction of the turn. If rudder is applied at this point, as it would be through normal pilot reaction to stop the spin, and the stick is not neutralized the airplane changes direction and spins toward the deflected rudder. A snap roll and a spin are the same thing

Above, ejection seat headrest. Below, aft end of right-hand pilot's console and circuit breaker panel.

FJ-3 PILOT'S RIGHT-HAND CONSOLE

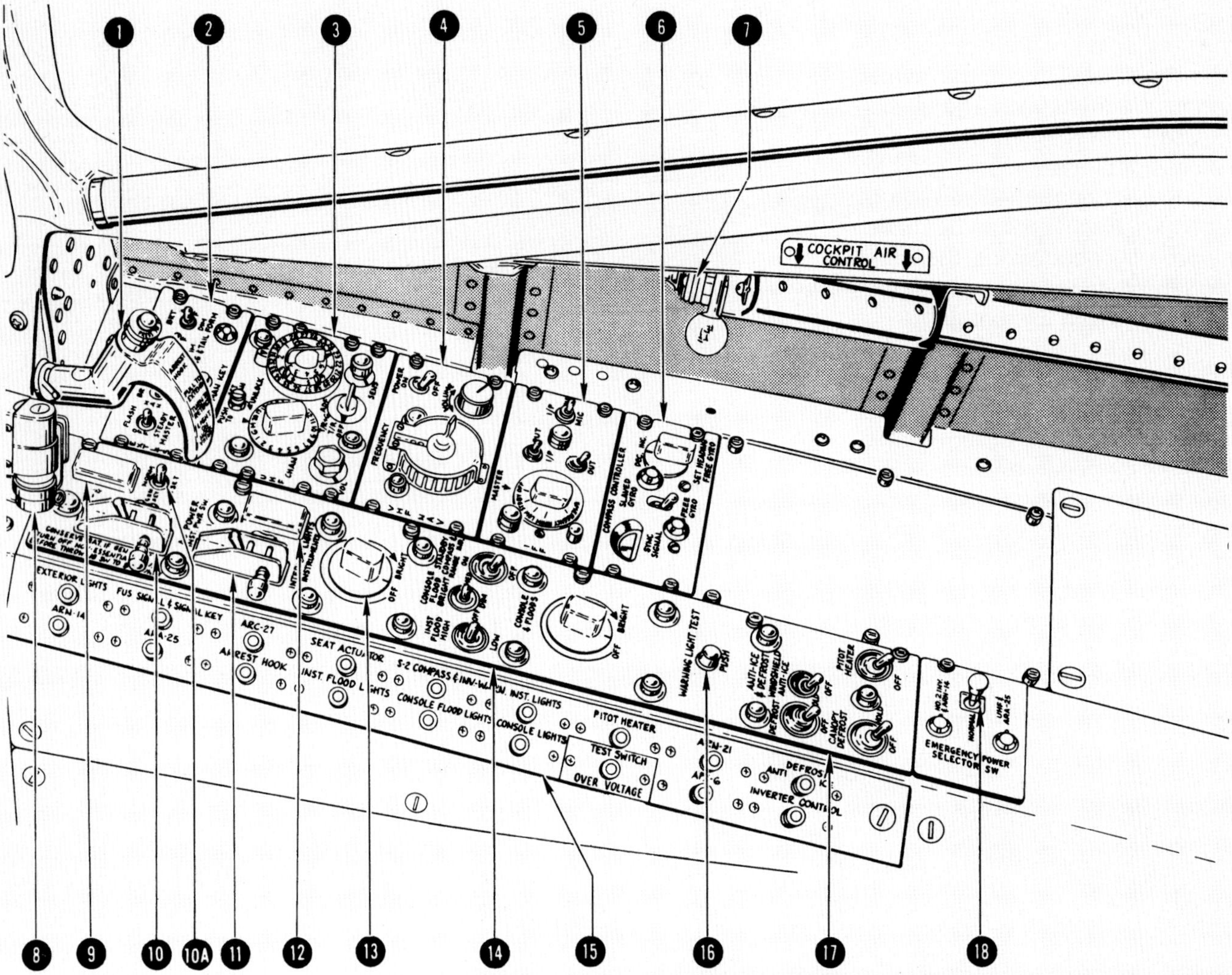

RIGHT-HAND CONSOLE

1. Arresting Hook Control
2. Exterior Lighting Control Panel
3. UHF Communications Control Panel
4. VHF Navigation Control Panel
5. IFF Control Panel
6. Gyrosyn Compass Control Panel
7. Cabin Air Deflector Control
8. Cockpit Light
9. Generator Warning Light
10. Battery and Generator Switch

10A. Generator Reset Switch †

11. Instrument Power Switch
12. Inverter Warning Light
13. Interior Light Rheostat
14. Interior Light Control Panel
15. Circuit - breaker Panel
16. Warning Light Test Switch
17. Anti - ice and Defrost Panel
18. Emergency Power Selector Switch*

*AIRPLANES HAVING SERVICE CHANGE NO. 381 COMPLIED WITH

† AIRPLANES HAVING SERVICE CHANGE NO. 374 COMPLIED WITH

except that the rotating axis of a snap roll is horizontal and that of a spin is vertical.

TWO TYPES OF SPINS

Two different types of spins can be encountered in the FJ-3. One type is a steep slow rotating, hesitating spin during which the airplane rotates at an average rate of 50° per second, loses about 2,000 ft of altitude clean or 2,700 ft altitude, tank-on, per turn, and recovery from which requires 1/4 to 3/4 of a turn after simultaneously placing the stabilizer to about 0° deflection and the rudder to a full anti-spin position.

The other type of spin encountered is a type that is much more oscillatory in nature (the nose rising as much as 45° above the horizon and dropping beyond the vertical in the first turns of an accelerated entry) and much steadier with respect to rotational velocity, during which the airplane rotates at an average rate of about 120° per second, loses about 1,600 ft of altitude tanks-on, per turn. Recovery requires from one to two turns simultaneously placing the stabilizer to about 0° deflection and the rudder to the full anti-spin position.

A marked increase in rotational velocity occurs after the controls are placed in the recovery positions during this second type of spin. The number of turns required to affect recovery after the application of recovery controls depends on the presence of external tanks and the amount of fuel carried in them. The airplane will recover with tanks-on and empty in 1 to 1.5 turns and with tanks-on and full in 2 turns from a fully developed spin.

FAST AT THE LAST

It is important to note that spin recovery is progressing satisfactorily when the spin is noted to speed up markedly after the application of recovery controls. Therefore, it must be remembered that controls should not be moved during spin recovery simply because the rate of spin rotation has rapidly increased. This is a

6-3 WING AND WING FENCE

Wing fence development for the 6-3 wing in March 1955. Above, the short fence outboard of the fuel tank was very near the final design. Note the two mini-fences tested inboard of the drop tank. (NAA)

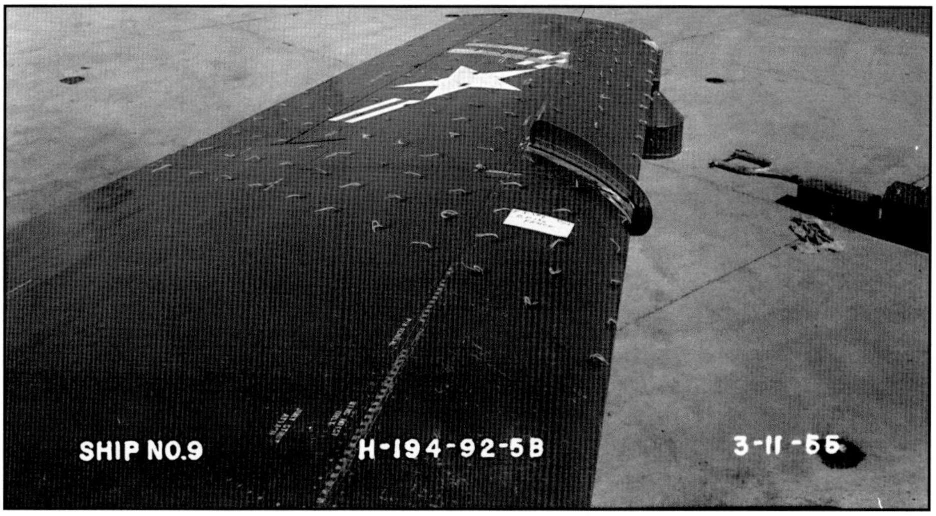

FJ-3M WING FENCES

good sign and an indication that recovery will occur within one to two turns. Do not move the stick full forward during spin recovery as this will cause a very abrupt negative acceleration and force the airplane to a somewhat greater nose-down position than vertical. This causes excessive altitude loss during recovery.

Remember then, that if a spin is encountered in an FJ, recovery can be affected regardless of whether the airplane is flying flaps up, flaps down, speed brakes open, tanks on or off by simply neutralizing the stick and applying hard anti-spin rudder. It should be noted here that the ailerons must be neutralized to affect satisfactory spin recovery with external tanks installed."

At left, FJ-3M wing fences . (Ginter)

SPEED BRAKES

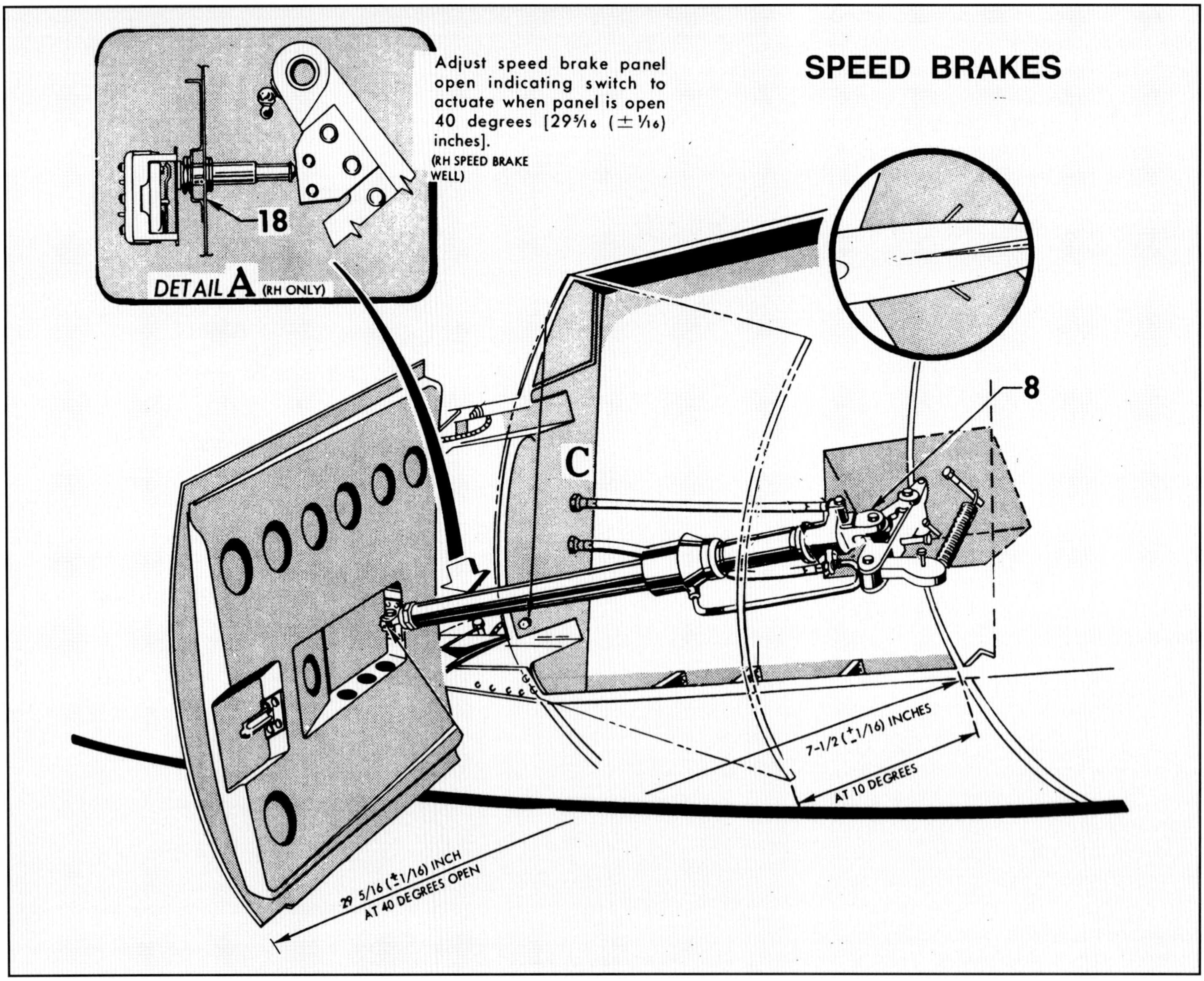

FJ-3 ARRESTING GEAR

Above and at right, FJ-3 tailhook prior to fitting of arresting hook doors. (National Archives)

The arresting gear control handle was on the inboard face of the right forward console. To extend the arresting hook, you pulled the handle aft until the pawl on the handle engages the slot at the hook down position. Movement of the handle aft to the hook down position unlocks the hook mechanical uplock and actuates the hydraulic selector valve which permits pressure from the arresting gear accumulator to lower the hook. When the handle was pulled aft, the barrier guard and barrier pickup were released. The hook was retracted by rotating the handle counterclockwise to release the pawl from the slot, allowing the handle to return to the spring-loaded hook up position. When the handle was returned to this position, the selector valve was actuated and the utility hydraulic system pressure was directed to the upside of the hook actuating cylinder.

A warning light was located on the arresting gear handle, illuminated whenever the hook was in a position that does not correspond to handle position.

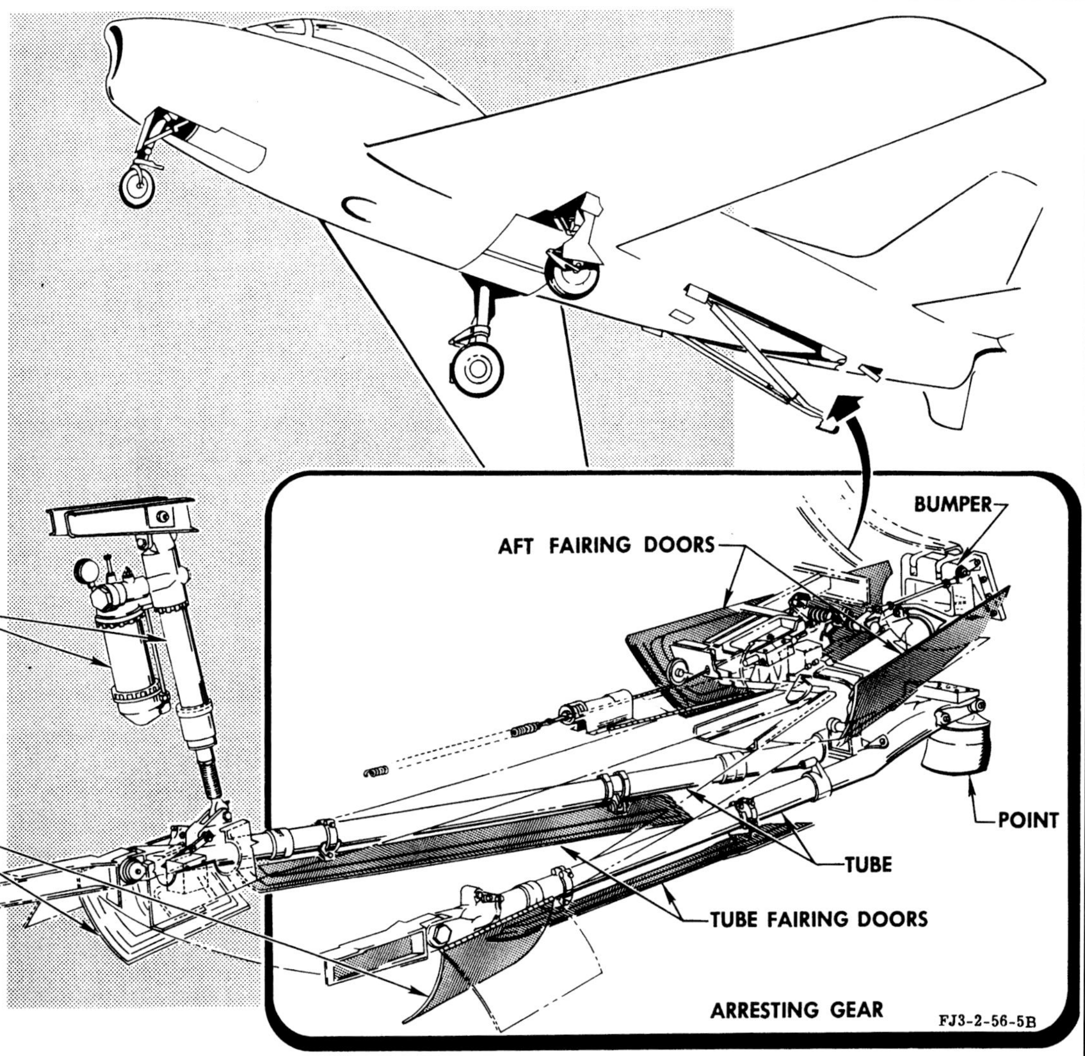

FJ-3 NOSE LANDING GEAR

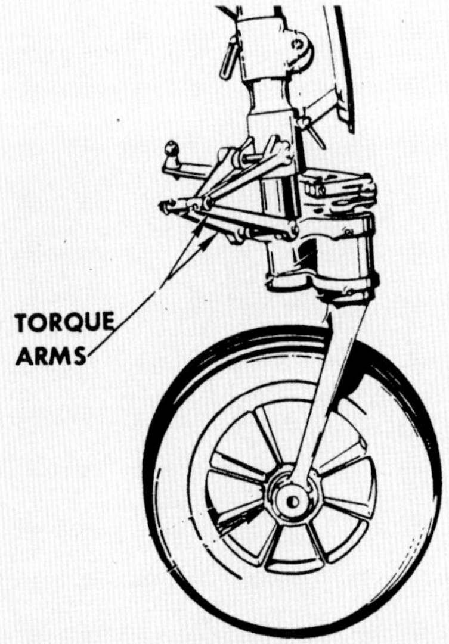

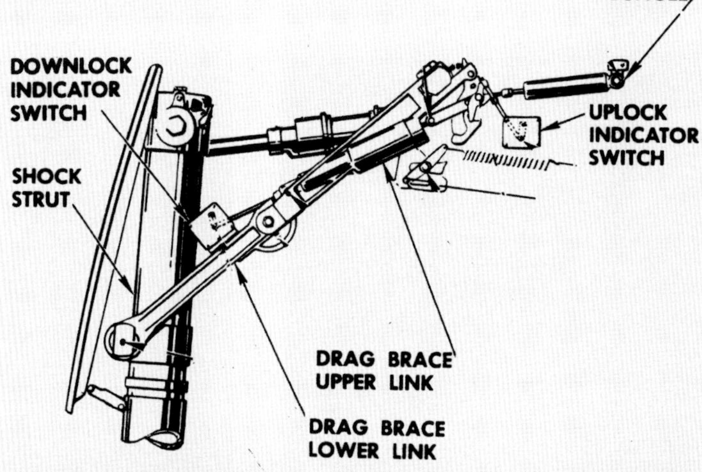

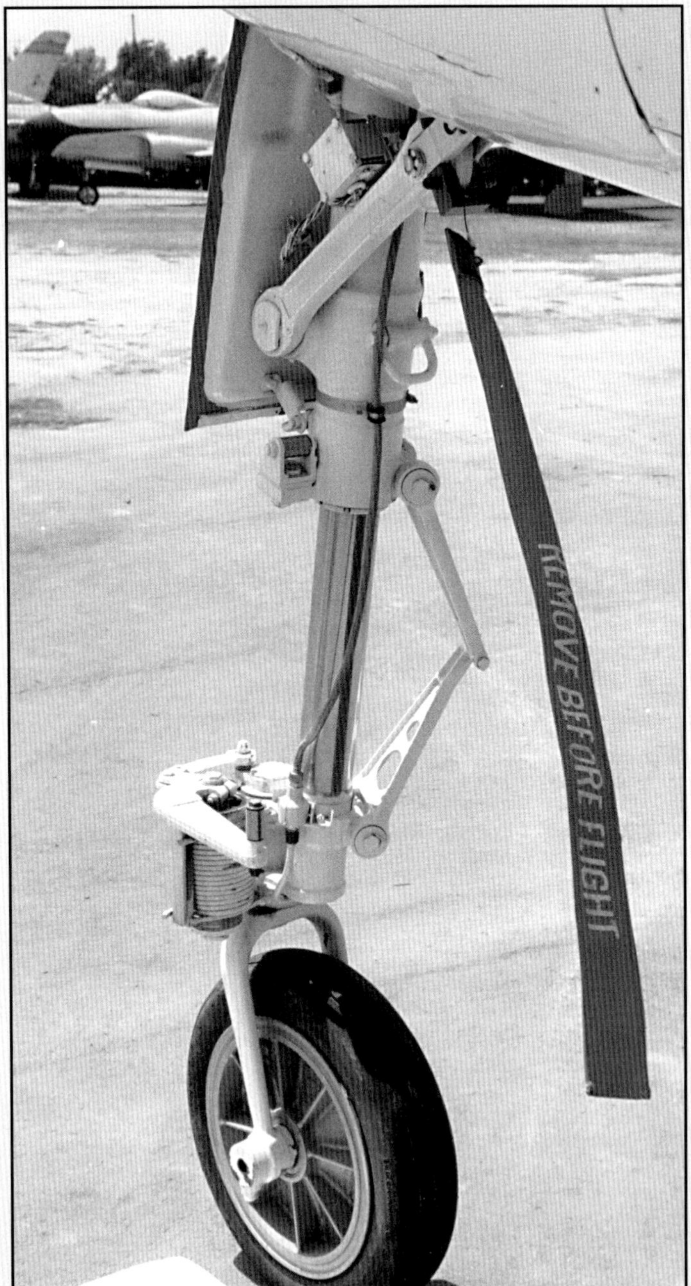

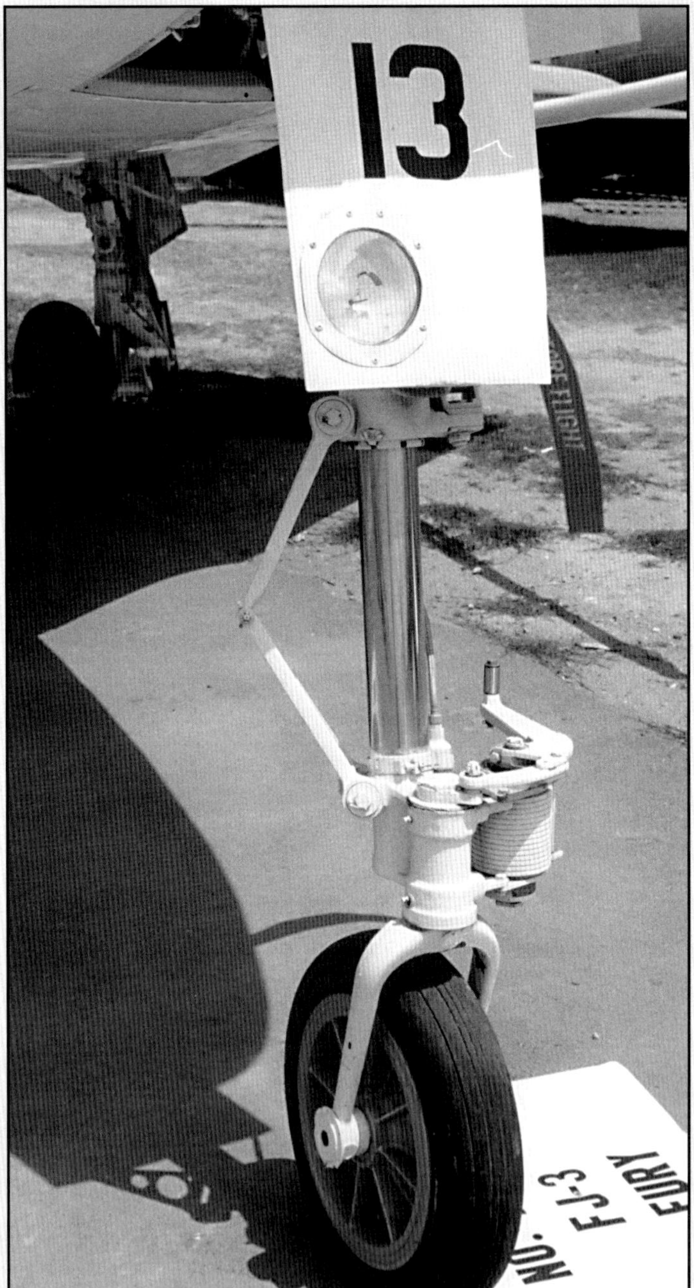

FJ-3 TAIL BUMPER

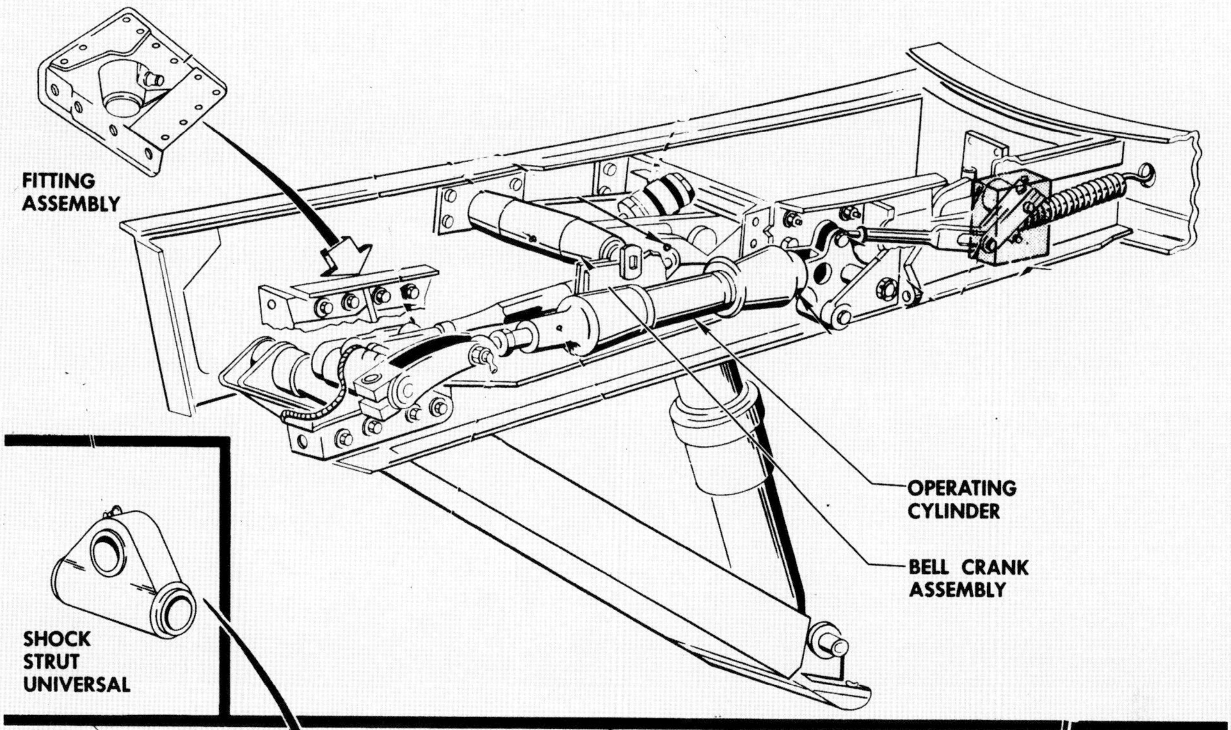

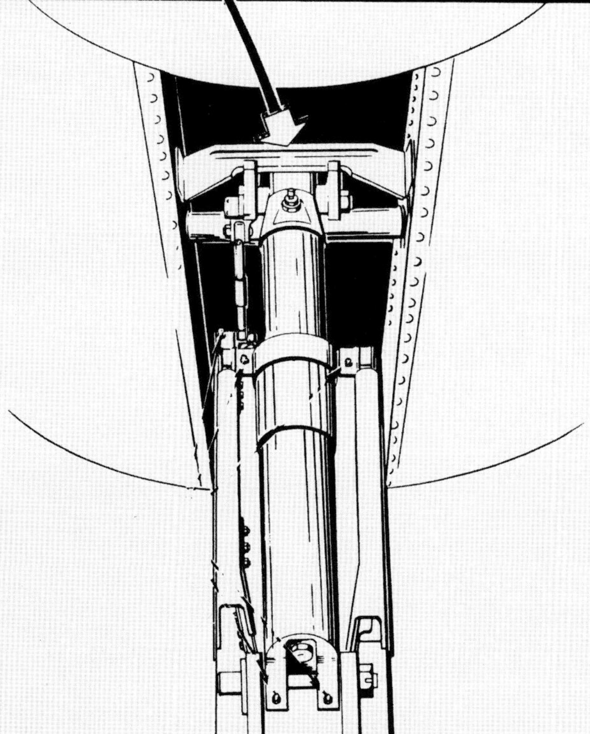

VIEW LOOKING FORWARD

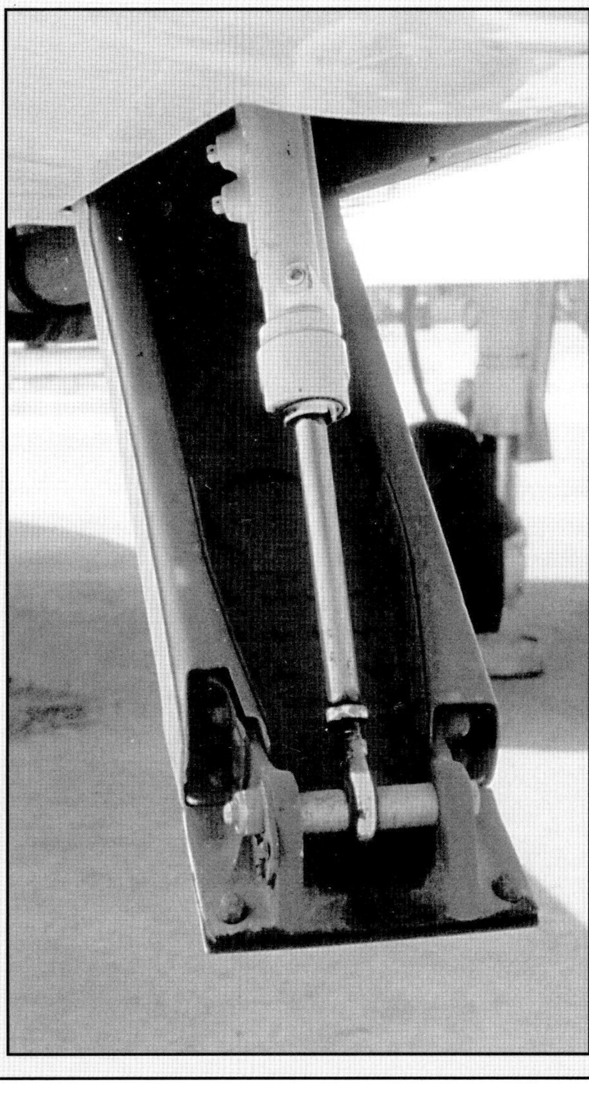

The hydraulically actuated tail bumper was extended and retracted in conjunction with the landing gear.

FJ-3 MAIN LANDING GEAR

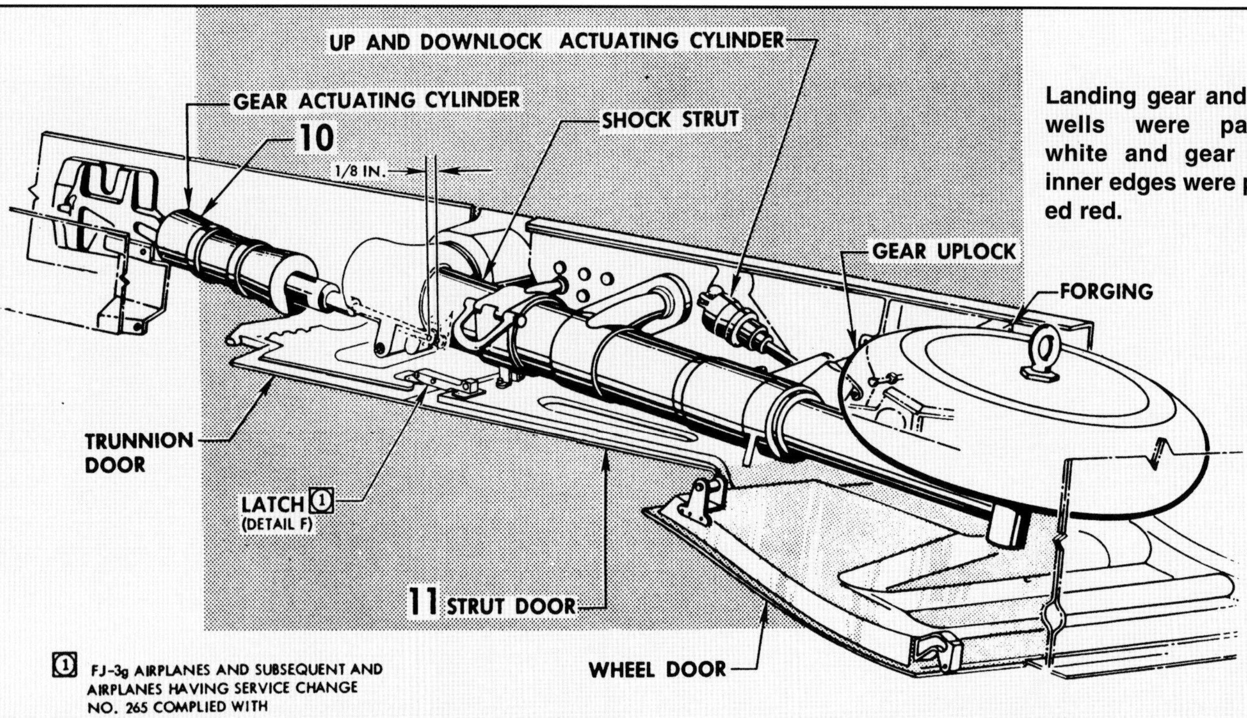

Landing gear and gear wells were painted white and gear door inner edges were painted red.

FJ-3 MAIN GEAR

The fully retractable tricycle landing gear and the gear and wheel fairing doors are hydraulically actuated and electrically controlled and sequenced. The main gear retracts inboard, into the lower surface of the wing and fuselage; the nose gear retracts aft, into the fuselage, pivoting 90°, so that the nose wheel is horizontal when retracted. After the gear is down and locked, the wheel fairing doors are retracted to the closed position reducing drag and preventing mud, dirt, etc., from entering the wheel wells during ground operation. The nonsteerable nose wheel makes use of a shimmy damper spring centering unit which permits rotation of the nose wheel through 360° of travel. The unit is so designed that it will return the nose wheel to the neutral position from any point within 70° right or left position.

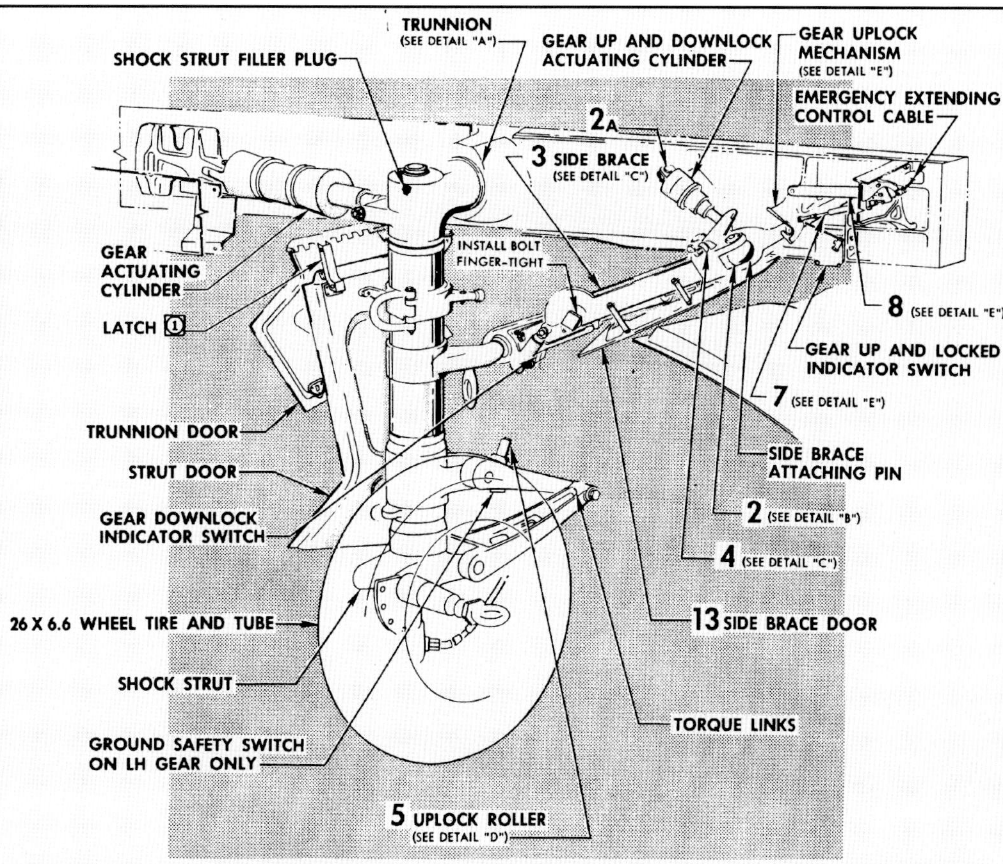

FJ-3 CATAPULT GEAR

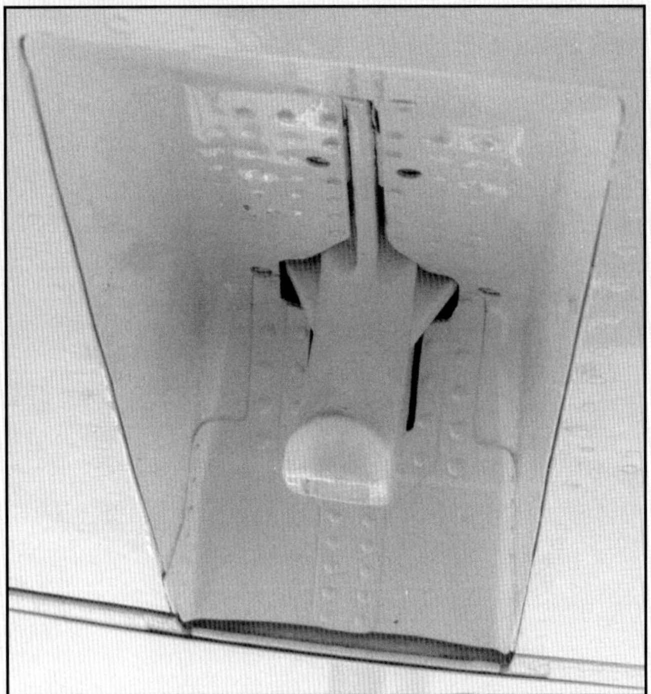

The catapult system consists of a pivoted catapult hook attached at the forward lower centerline of the fuselage and a holdback mechanism located on the outer lower centerline of the fuselage just aft of the fuselage break. The pendant (NAF314261-1) is attached to the catapult with the hook in the normal position. As the catapult is tensioned and the airplane catapulted, the hook lines up with the load path and returns to the faired position by bungee action when the pendant is released. The "coupon" (release element, tension bar type NAF502822-2) is installed in the holdback mechanism and hooked to the deck holdback and release cleat link unit (NAF500637-12); tension force keeps the airplane holdback open. When the "coupon" breaks, the holdback mechanism automatically retracts and the forward end of the "coupon" is retained in the airplane holdback mechanism.

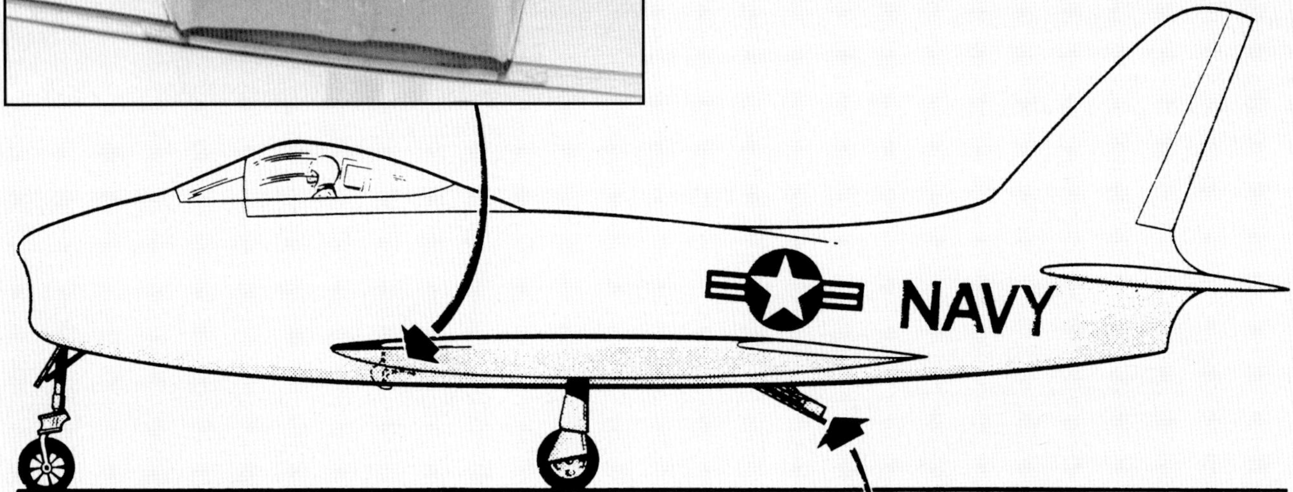

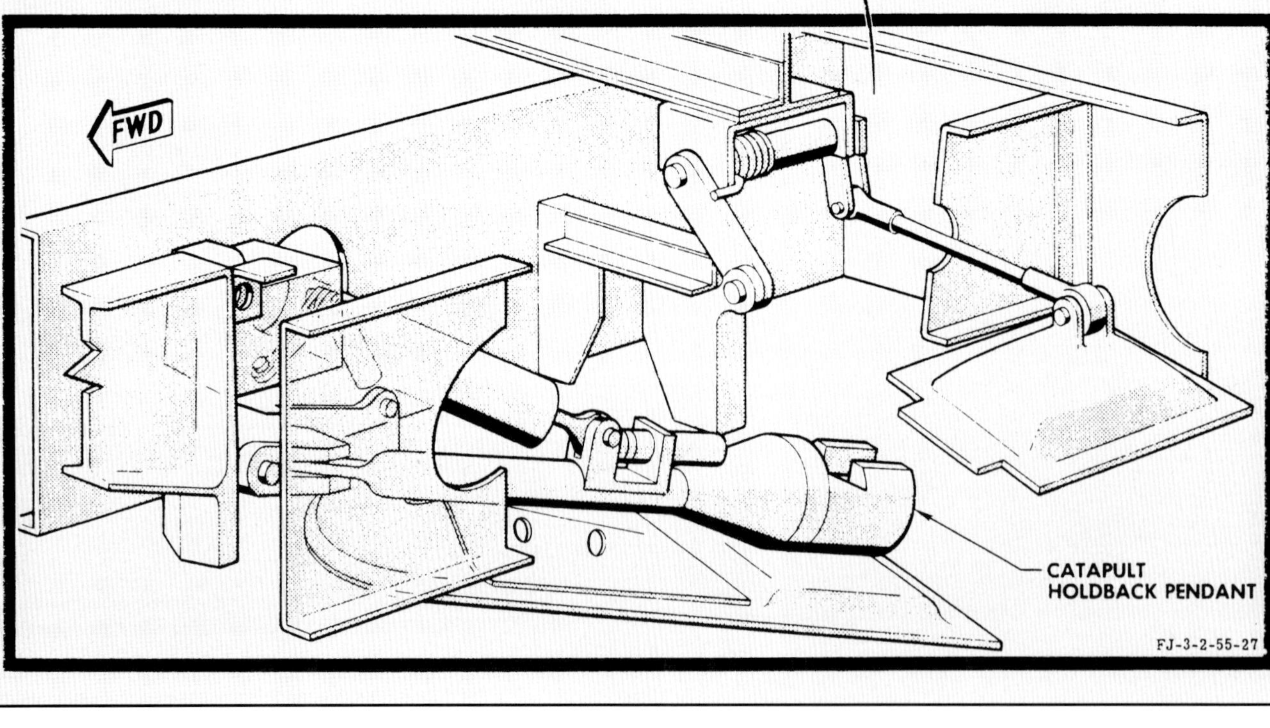

FJ-3 ARMAMENT CONTROL SYSTEM

The FJ-3 armament control system consisted of the AN/APG-30 radar set and the aircraft fire control system. The Mk. 11 Mod 1 sight unit and related components make up the aircraft fire control system. The armament control system automatically computes leads and produces a visual sighting image for use as a reference during target tracking.

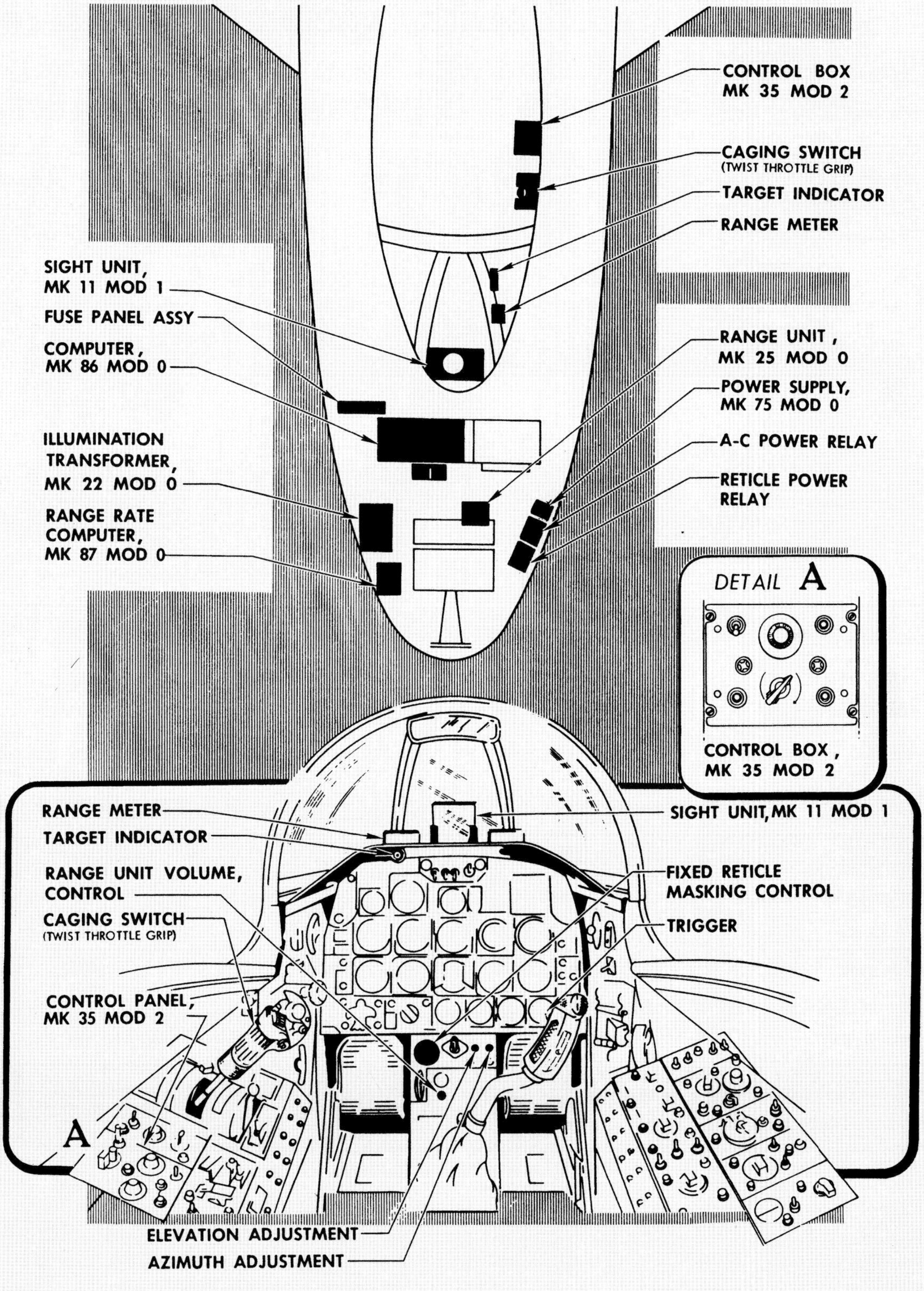

FJ-3 RADAR EQUIPMENT

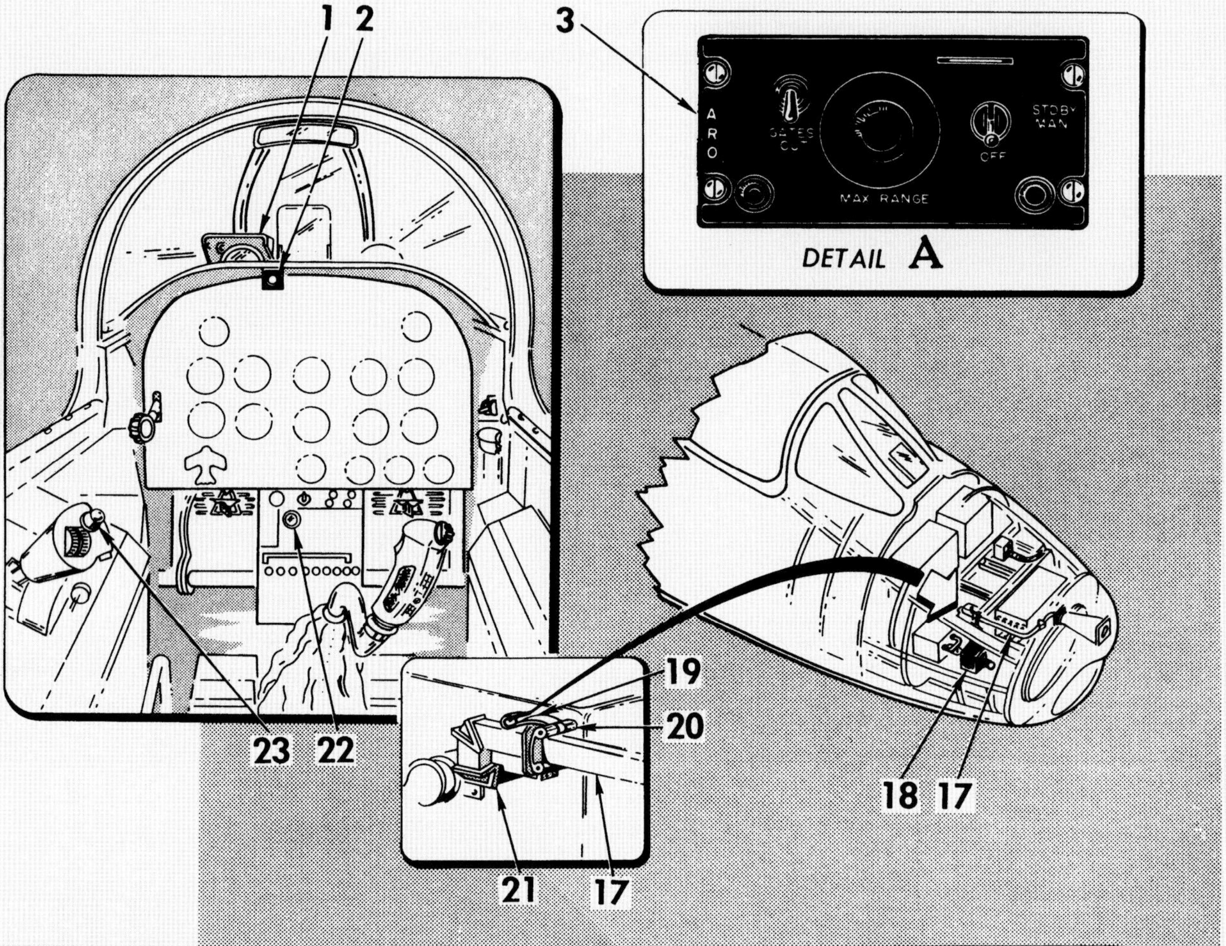

No.	Item	Part No.
1	RANGE METER	181-74222
2	TARGET INDICATOR	AN3157-2
3	CONTROL PANEL	C-775/APG-30
4	POWER SUPPLY—RANGE COMPUTER	PP-493/APG-30 [1] OR PP-493A/APG-30A [2]
5	FREQUENCY CONVERTER-TRANSMITTER	RT-181/APG-30 [1] OR RT-322/APG-30A [2]
6	IFF CONTROL PANEL	C-1159/APX-6B
7	RADAR SET CONTROL	C-1272/APA-89
8	FUSE PANEL	194-74150
9	TEST RECEPTACLE	AN3100-28-11S
10	TEST RELAY	AN3304-1
11	IFF ANTENNA	AT-234/APX-6 [3] 1305.126 (DORNE & MARGOLIN) [4]
12	IFF RECEIVER-TRANSMITTER	RT-82A/APX-6
13	A-C POWER SUPPLY RELAY	AN 3312-1
14	VIDEO CODER	KY-81/APA-89
15	VOLTAGE REGULATOR	CN-112/APG-30
16	ANTENNA	AT-223/APG-30 [1] OR AT-561/APG [2]
17	WAVE GUIDE	4732 (COOPERATIVE INDUSTRIES)
18	RANGE RATE COMPUTER	MARK 87 MOD 0
19	WAVE GUIDE LATCH ASSEMBLY	194-74035
20	WAVE GUIDE HINGE ASSEMBLY	194-74032
21	WAVE GUIDE DIRECTIONAL COUPLER	66725 (WAVELINE INC.) [1] 64015 (WAVELINE INC.) [2]
22	CRYSTAL CURRENT METER	194-54226
23	RANGE GATE SWITCH	138106 (UCINITE)

[1] AIRPLANES 135774 THROUGH 139278 NOT HAVING SERVICE CHANGE NO. 438 COMPLIED WITH

[2] AIRPLANES 141364 AND SUBSEQUENT AND AIRPLANES HAVING SERVICE CHANGE NO. 438 COMPLIED WITH

[3] AIRPLANES 135774 THROUGH 136142

[4] AIRPLANES 136143 AND SUBSEQUENT

FJ-3-2-74-11C

FJ-3 RADAR EQUIPMENT

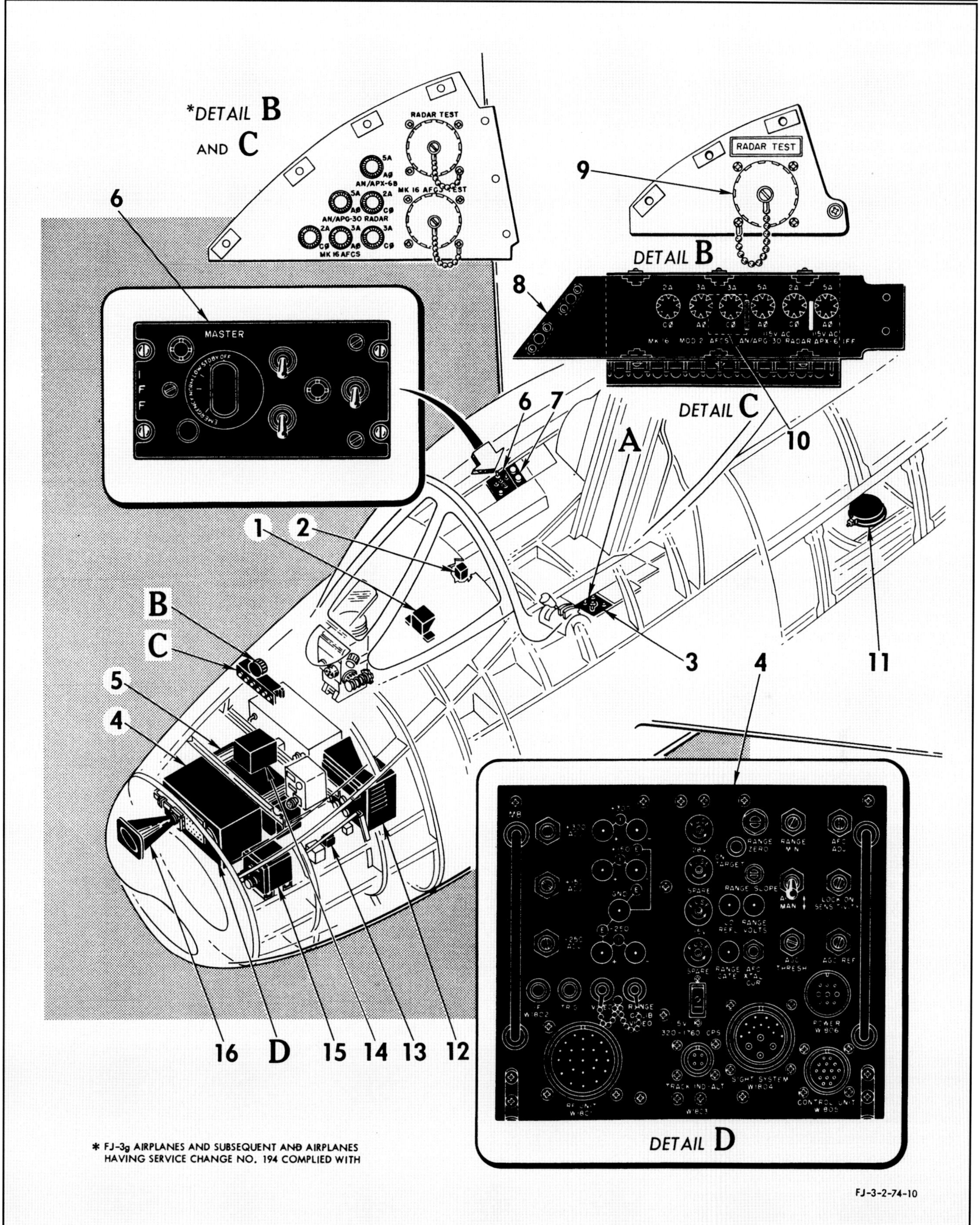

* FJ-3g AIRPLANES AND SUBSEQUENT AND AIRPLANES HAVING SERVICE CHANGE NO. 194 COMPLIED WITH

FLIGHT TESTING THE FJ-3

By Capt. Robert Dreesen USN (Ret)

Capt. Dreesen flight tested the FJ-3 at NATC and described it as a "lively, comfortable, very smooth-flying aircraft that could take almost anything else in the air at the time in a dogfight situation; it could fly rings around the F9F-8, for example. But it was a lousy aircraft for air-to-air combat because of a design innovation in the FJ-2 carried over the FJ-3. Someone in BuAer had decided that it would be a big advantage if the gun line were adjusted downward so that under a specified set of flight conditions, the gun line would correspond to the actual aiming point. The problem was that for all other combinations of flight conditions, the gun line was way below the roll axis. This made tracking a moving target very difficult. Testing the FJ-2 at Patuxent had established this problem, but by the time the report was in it was too late to make a change for the FJ-3."

Dressen related that in a dogfight situation, the pilot would have been limited to quick snap shots, with no tracking at all. Accurate shooting in the FJ-3 depended almost entirely on the pilot's eyes rather than on the gunsight. Dreesen cited Fleet Gunnery Reports that clearly supported the NATC report. "Every aircraft in the Fleet could outshoot the Fury against target banners."

The gunnery system of the airplane consisted of four Mk. 12 Mod 0 20mm cannons. These fixed forward-firing guns were mounted in pairs in each side of the nose section, below and forward of the cockpit. The guns were supported by trunnion yoke mounts at the forward end of the receiver and by adjustable trunnion mounts at the aft end of the receiver. Each gun was equipped with a Mk. 7 Mod 0 pneumatic feeder and a Mk. 1 Mod 1 synchronizer switch. Ammunition boost motors and muzzle stabilizers were provided for each gun. Each of the four ammunition containers carried 150 rounds of 20mm. Expended ammunition links and cases were retained in compartments provided in the airplane. The gun bay purging system purges explosive gases from the gun area during gun firing.

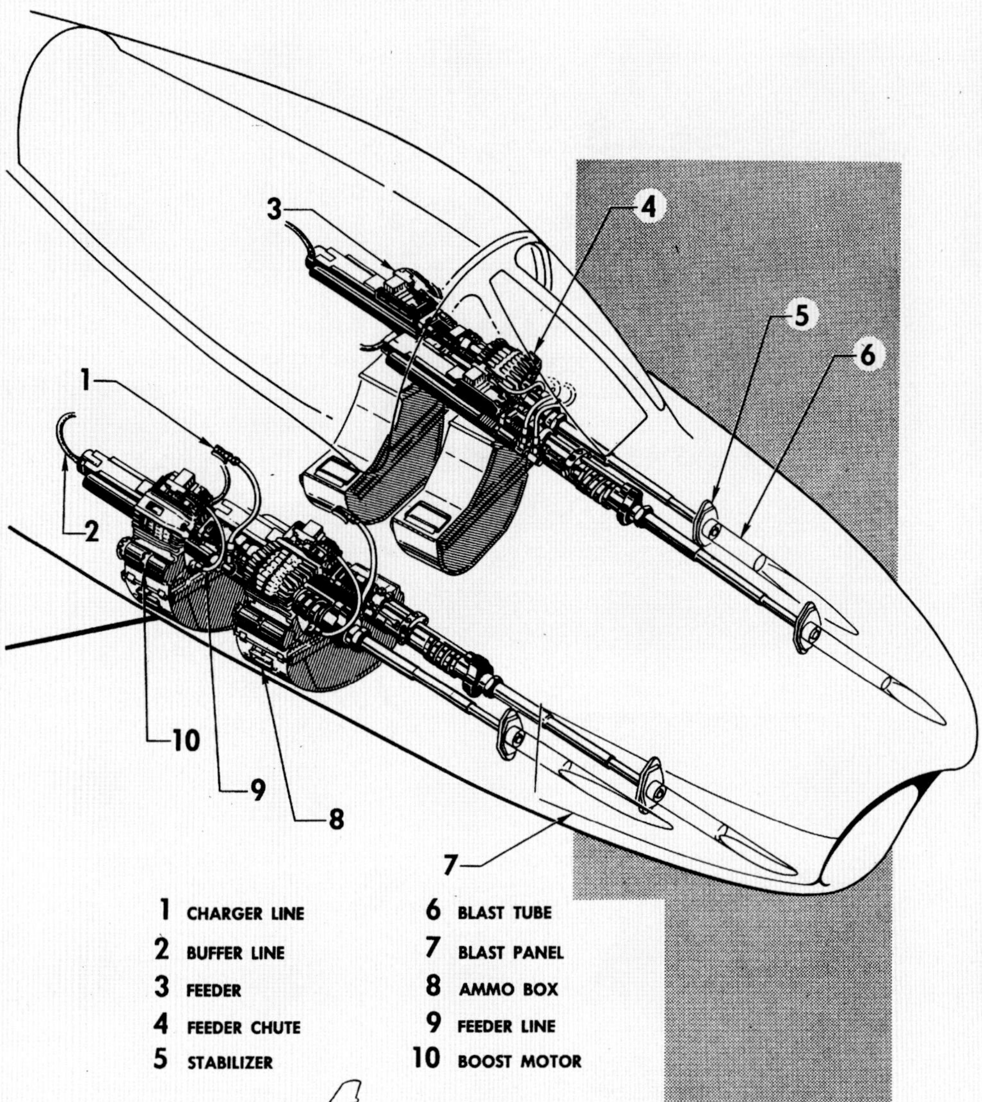

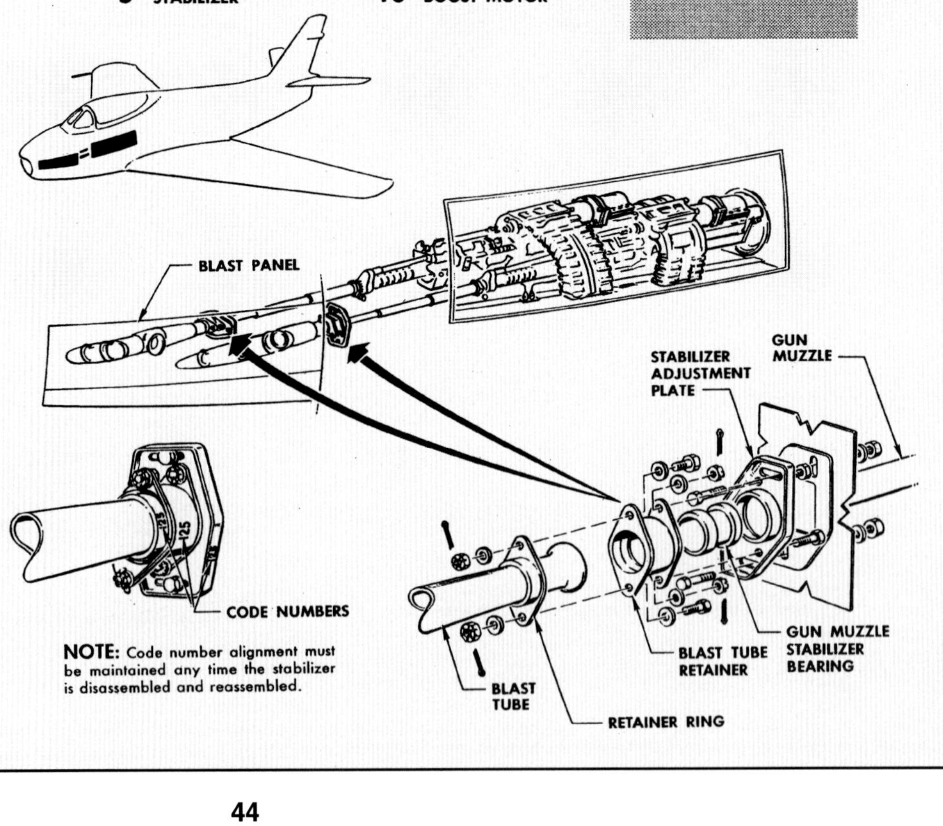

FJ-3 GUNNERY EQUIPMENT

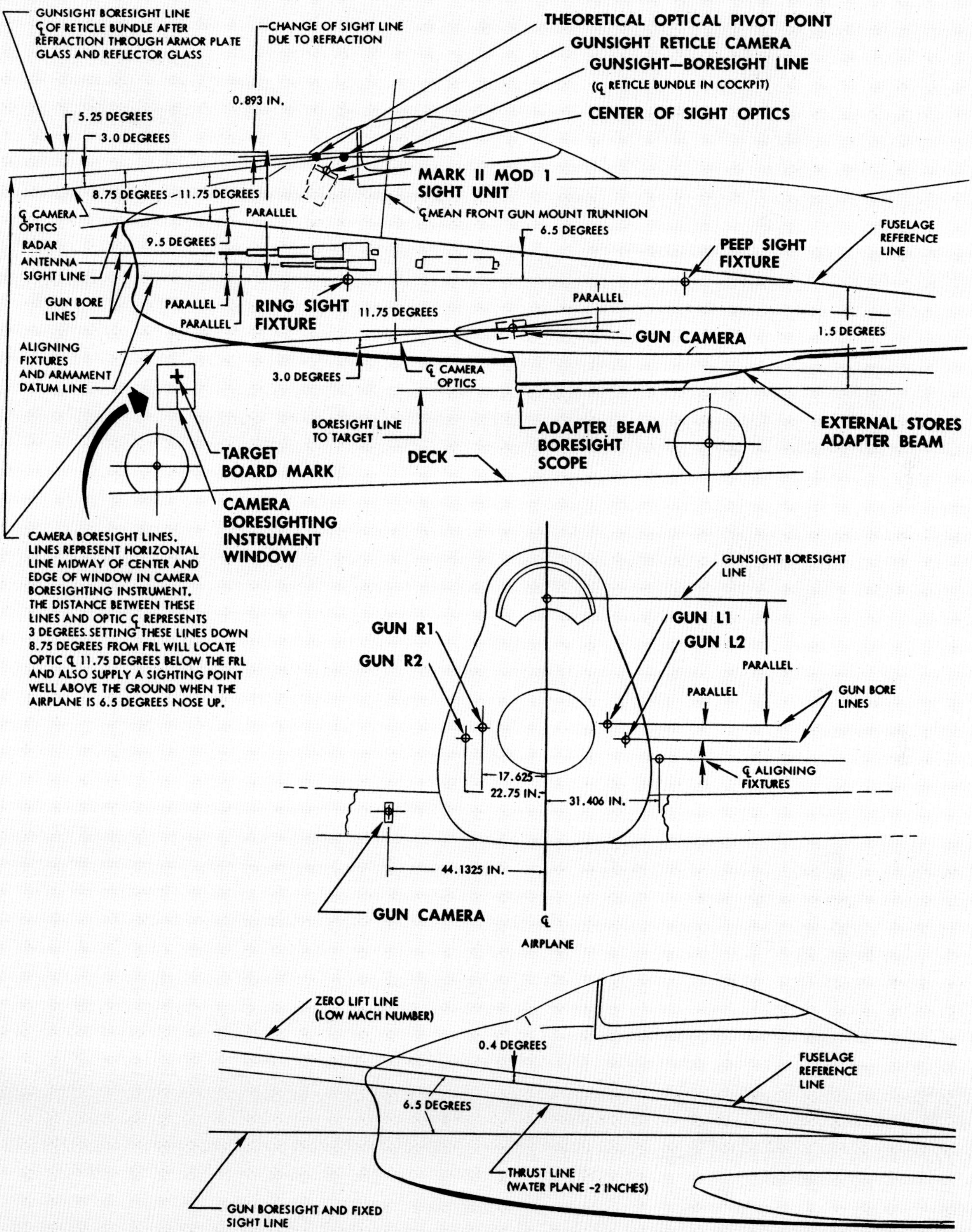

FJ-3 GUNNERY EQUIPMENT

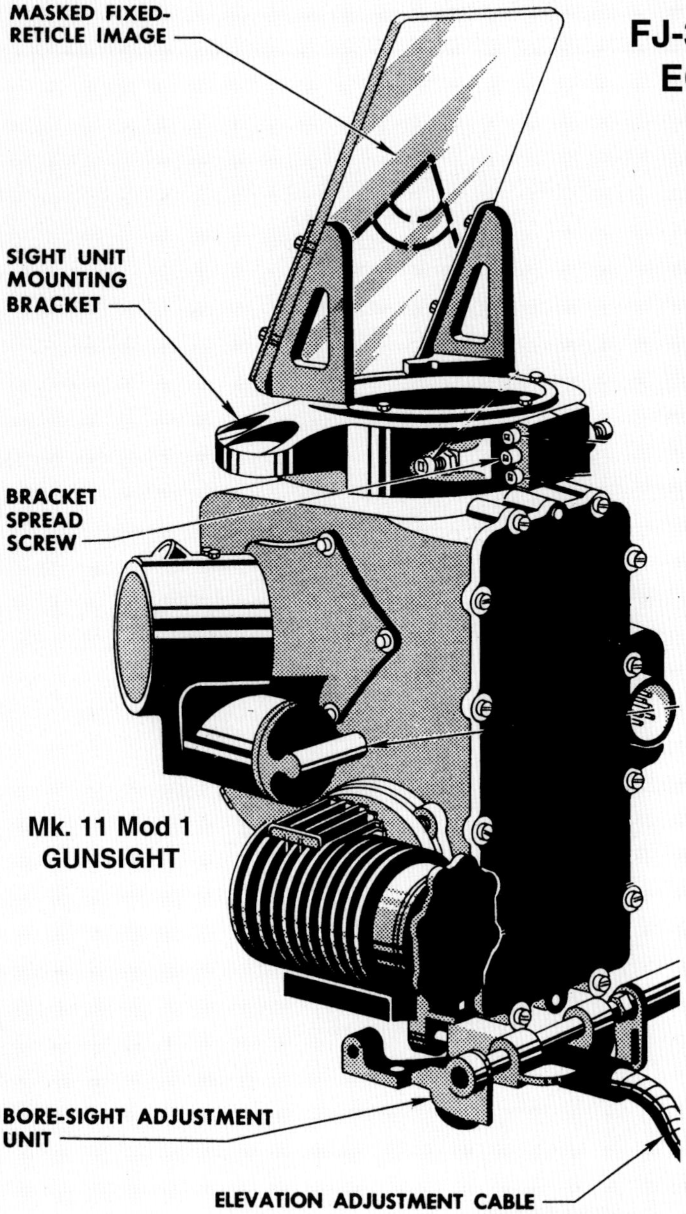

Mk. 11 Mod 1 GUNSIGHT

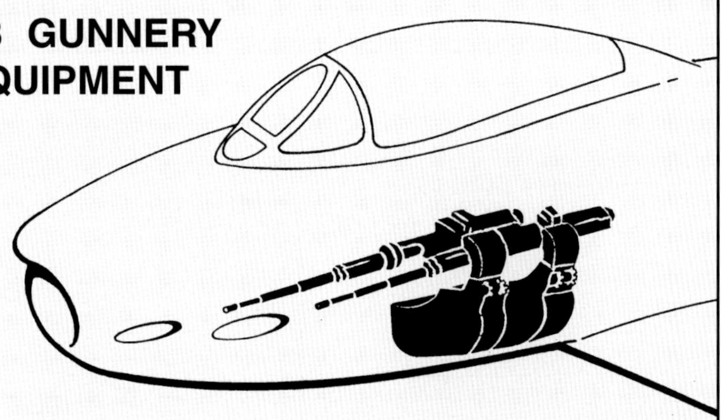

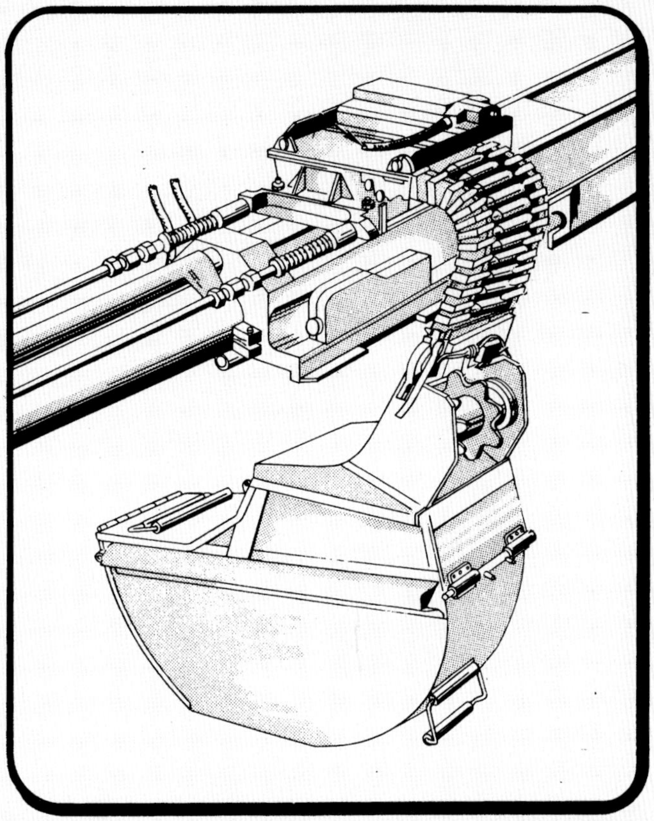

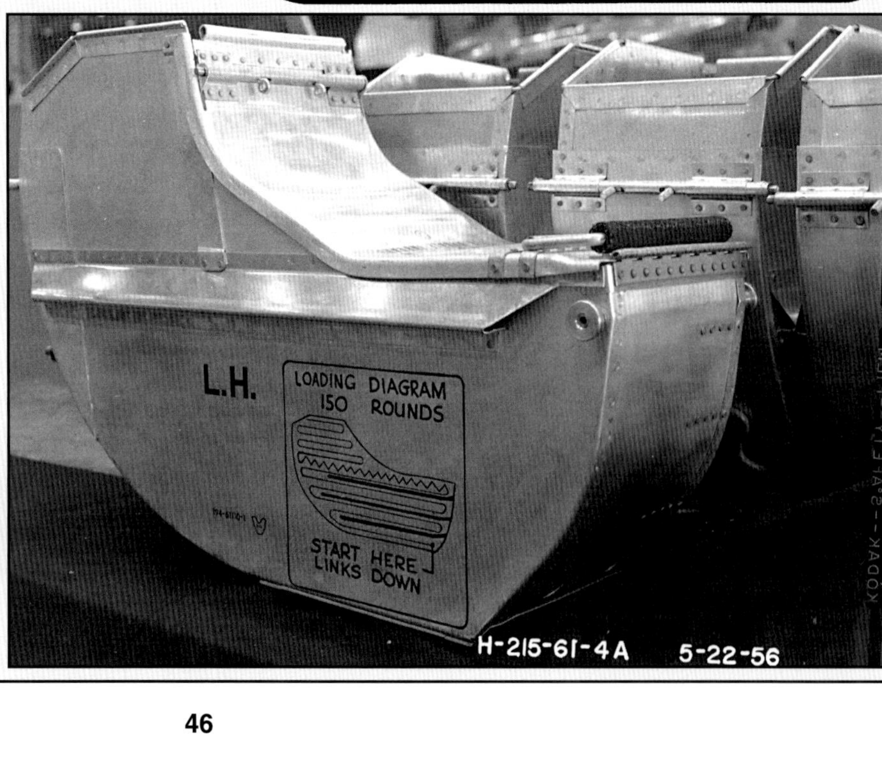

Above, ammo can door open.
At right, FJ-3 150 round ammo cans.

FJ-3 GUNNERY EQUIPMENT

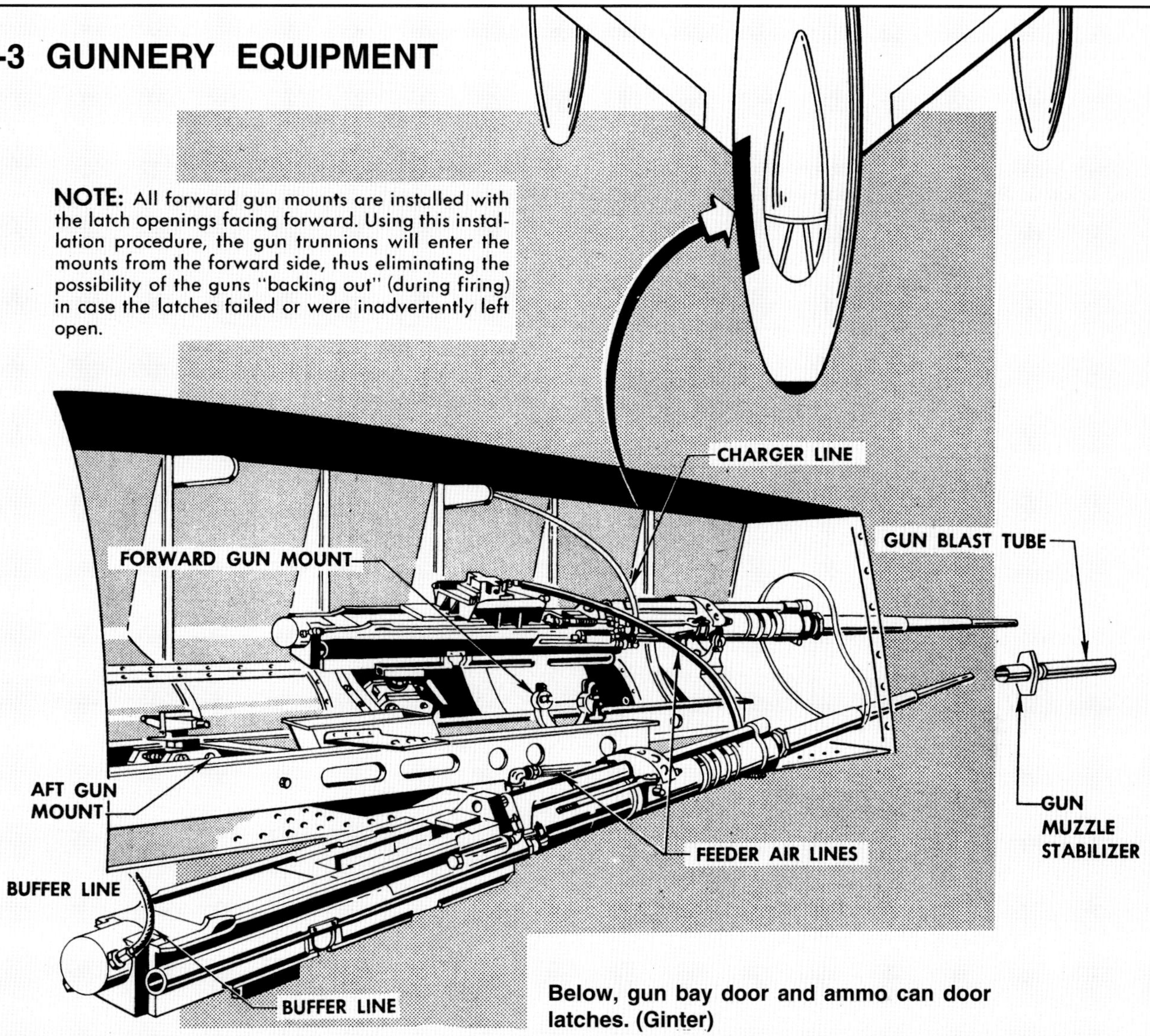

Below, gun bay door and ammo can door latches. (Ginter)

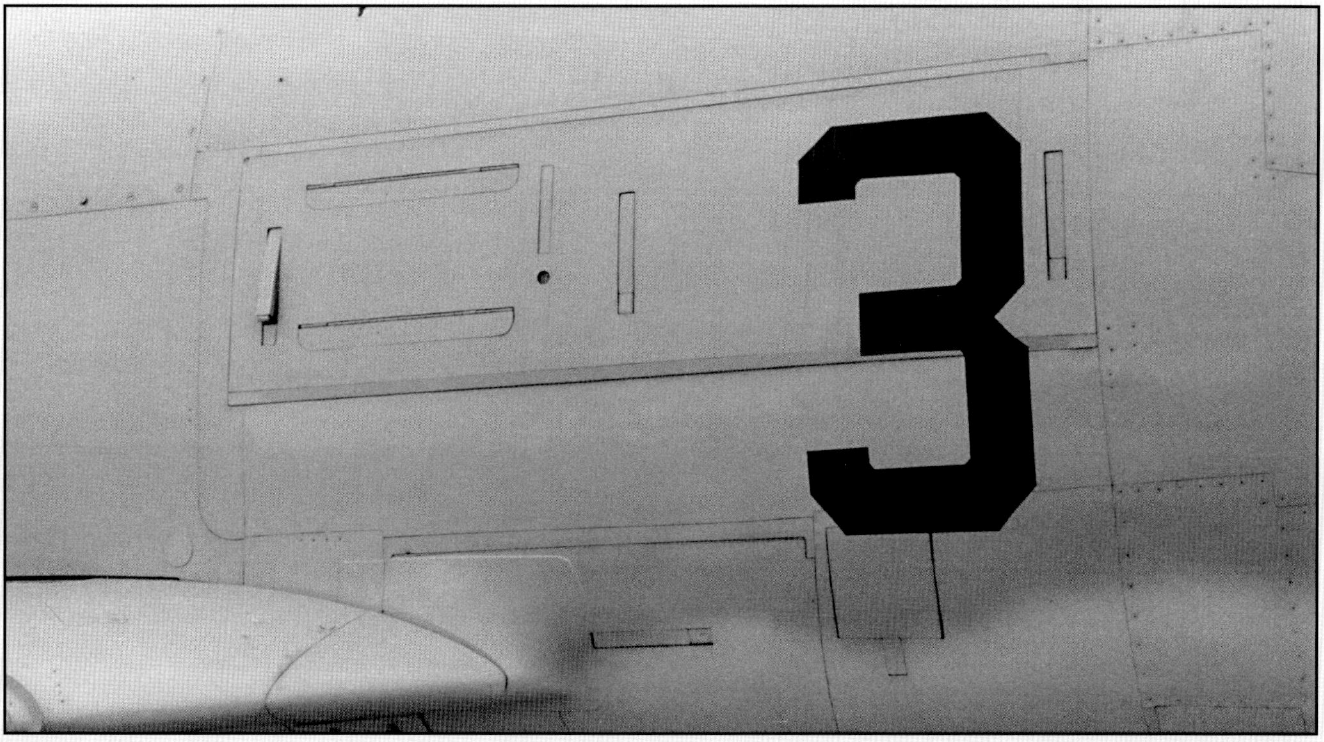

FJ-3/3M BOMBING EQUIPMENT BuNos 136118 and SUBSEQUENT

FJ-3 aircraft BuNos 136118 and subsequent were fitted with six hard points instead of two and fitted with a bomb/rocket control box (seen here) on the center pedestal. Most became FJ-3Ms with the addition of the wiring needed to support Sidewinder missiles.

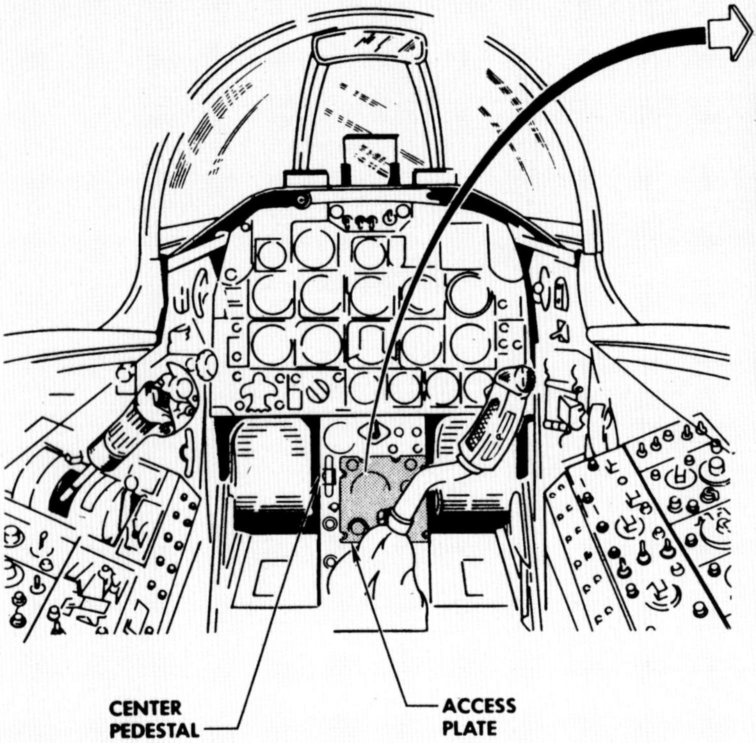

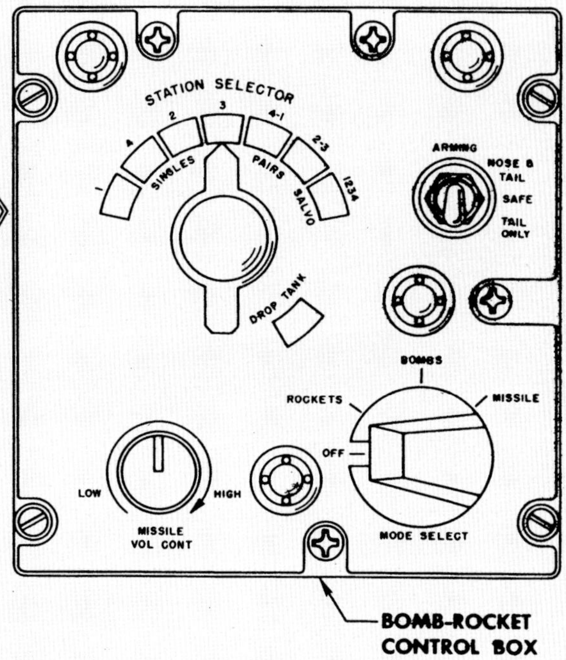

1 Remove access plate from face of center pedestal.

2 Route electrical connector plugs out through access opening.

3 Connect electrical connector plugs to the two receptacles on back of bomb-rocket control box.

4 Place bomb-rocket control box in mounting position and secure fasteners.

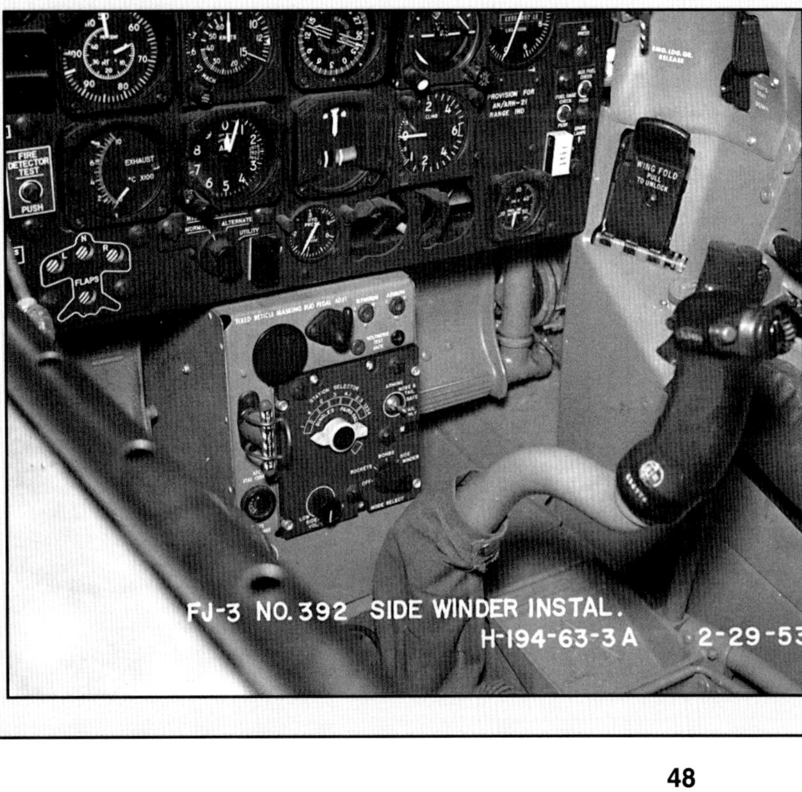

FJ-3M COCKPIT

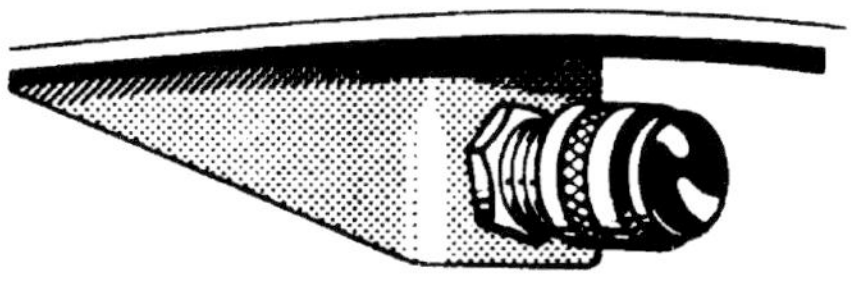

ON-TARGET INDICATOR LIGHT

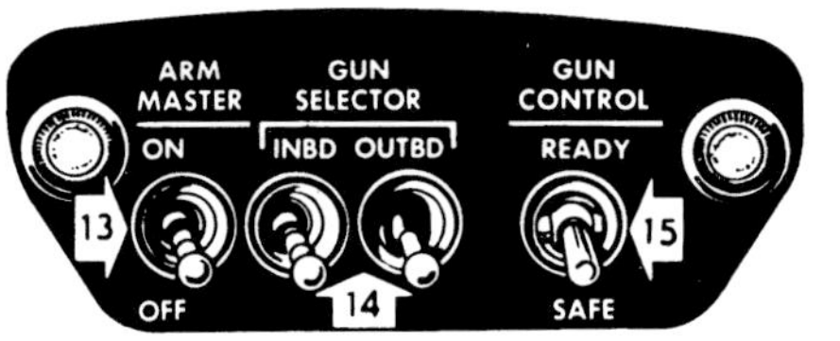

GUN CONTROL PANEL

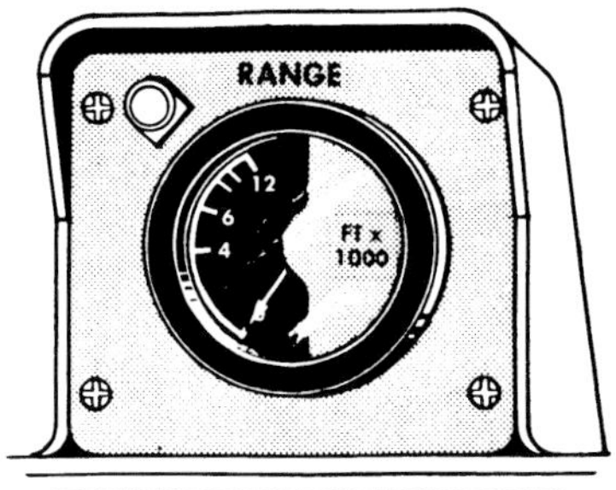

RANGE METER

THROTTLE

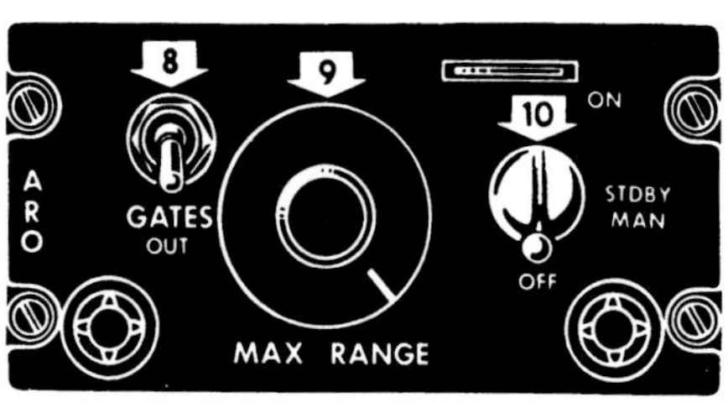

RADAR CONTROL PANEL

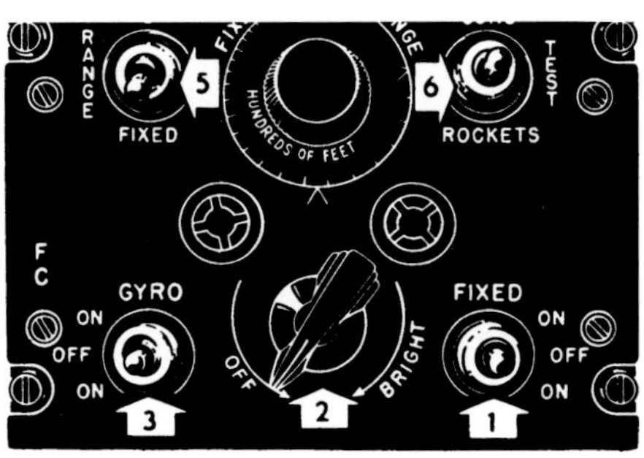

FIRE CONTROL PANEL

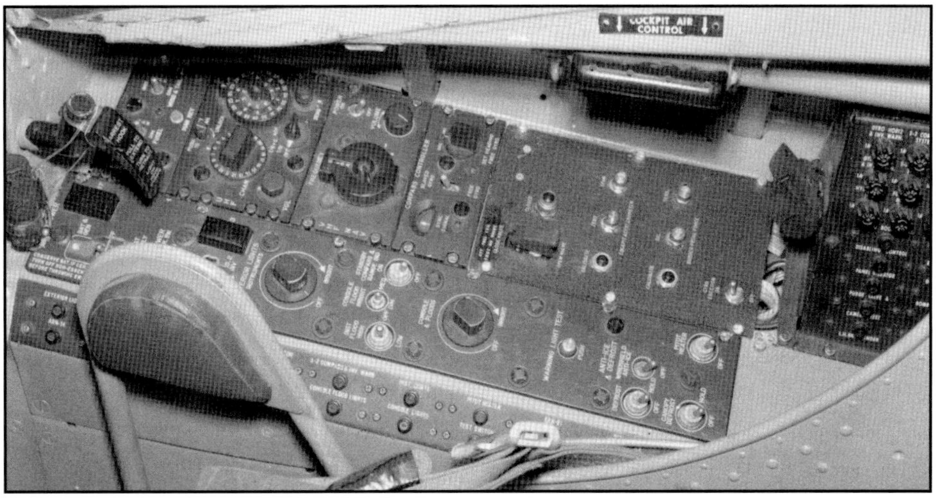

AERO 6A-1 and AERO 7 ROCKET PODS

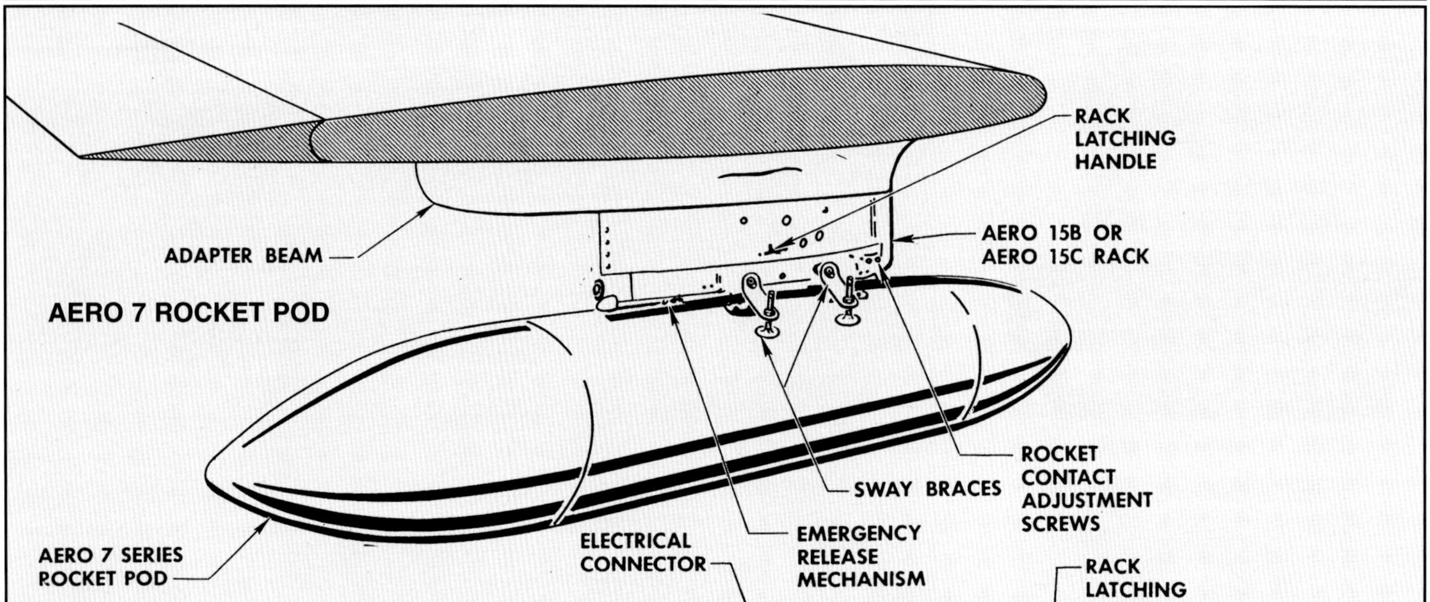

AERO 7 ROCKET POD

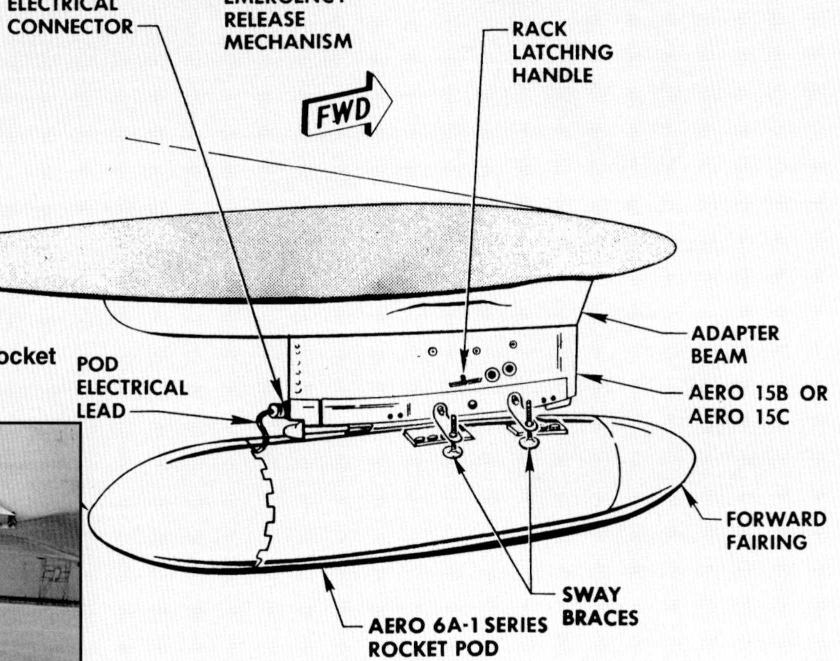

AERO 6A-1 ROCKET POD

Below, Aero 7 rocket pod (left) and Aero 6A-1 rocket pod (right) viewed from behind. (USN)

Below, head-on view of FJ-3 with Aero 6A-1 rocket pods on the inner pylons and Aero 7 rocket pods between the main landing gear and the wing drop tanks. (USN)

AERO 6A-1 and AERO 7 ROCKET PODS and PYLONS

At left and above, Aero 6A-1 and Aero 7 rocket pods on FJ-3M 462. (USN)

Below, single 5" HVAR mounted on Aero 15B/C weapons pylon.

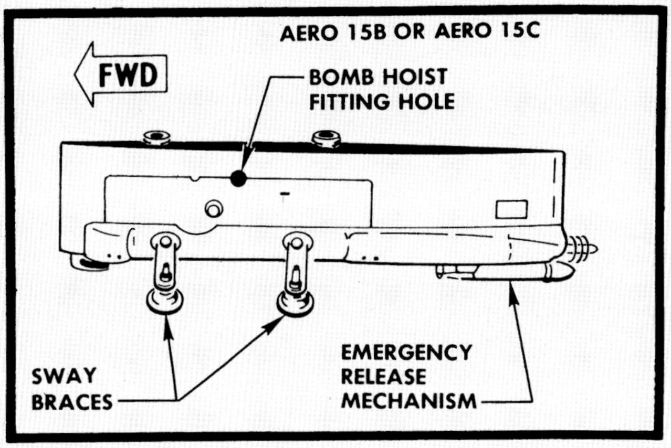

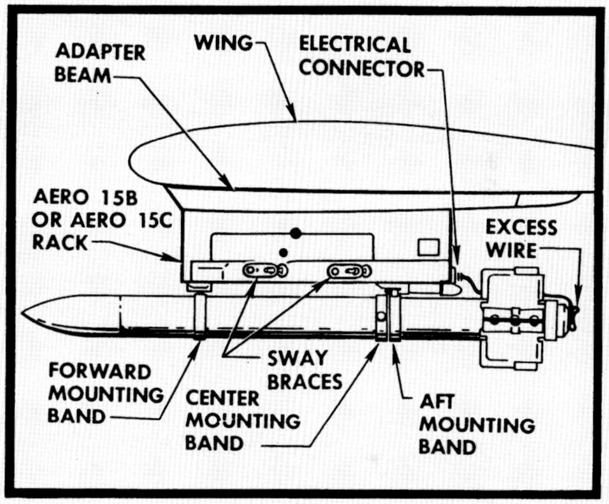

EXTERNAL STORES ADAPTER BEAM

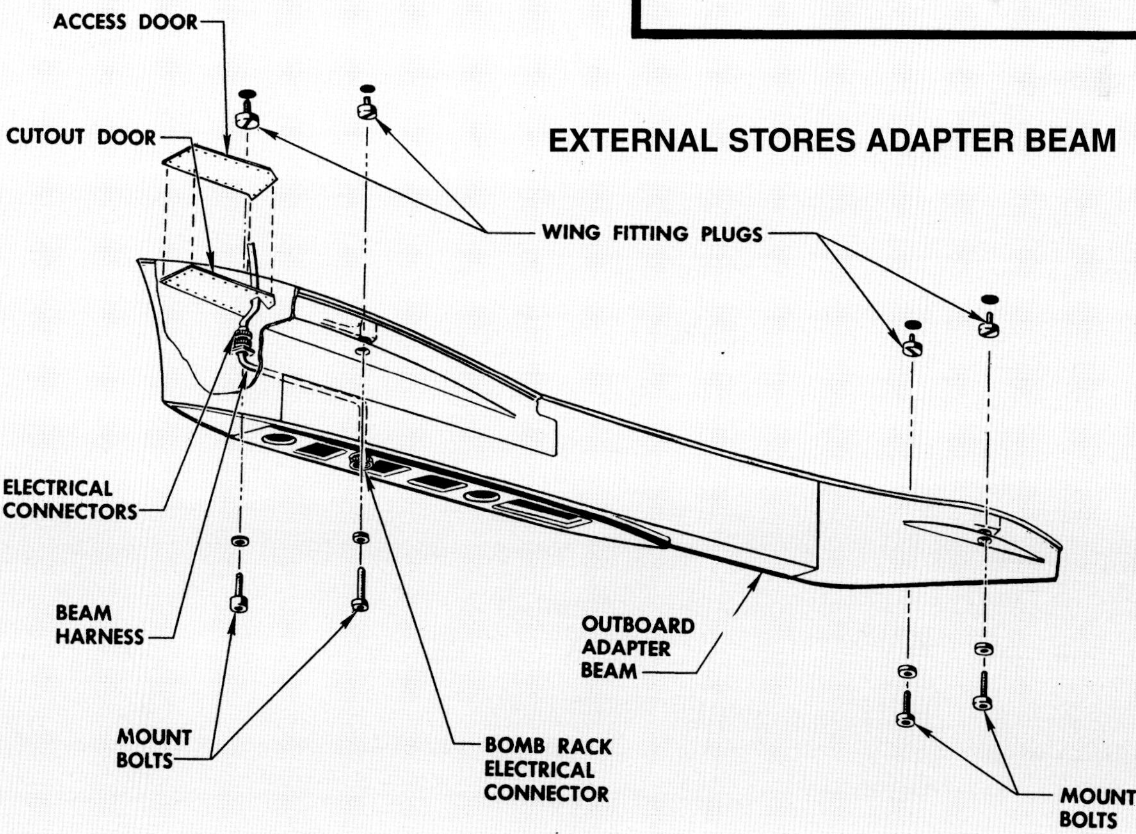

AERO 65A BOMB RACK and 1,000 POUND BOMB

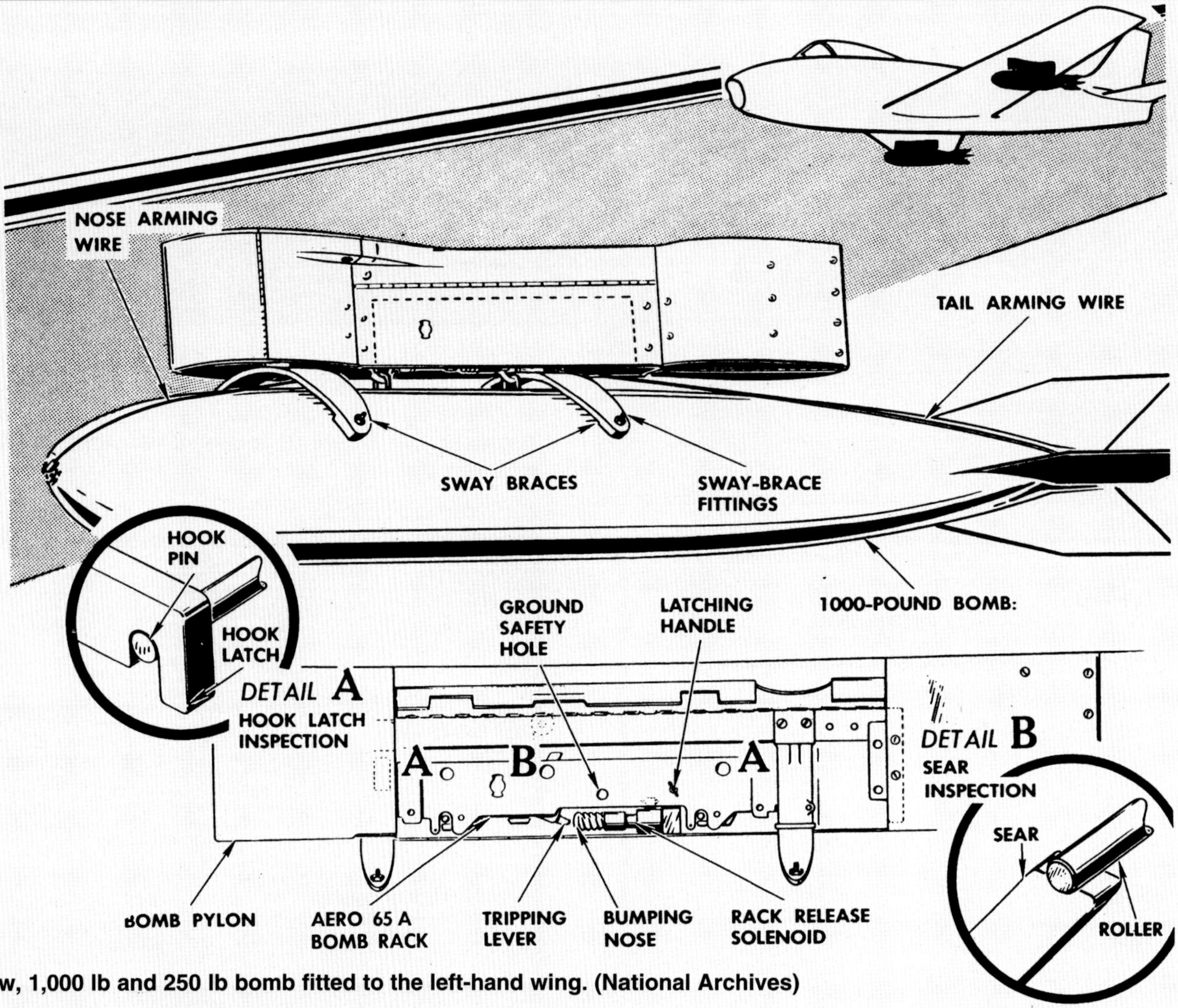

Below, 1,000 lb and 250 lb bomb fitted to the left-hand wing. (National Archives)

AERO 15A/B BOMB RACK and ADAPTER BEAM with 500 POUND BOMB

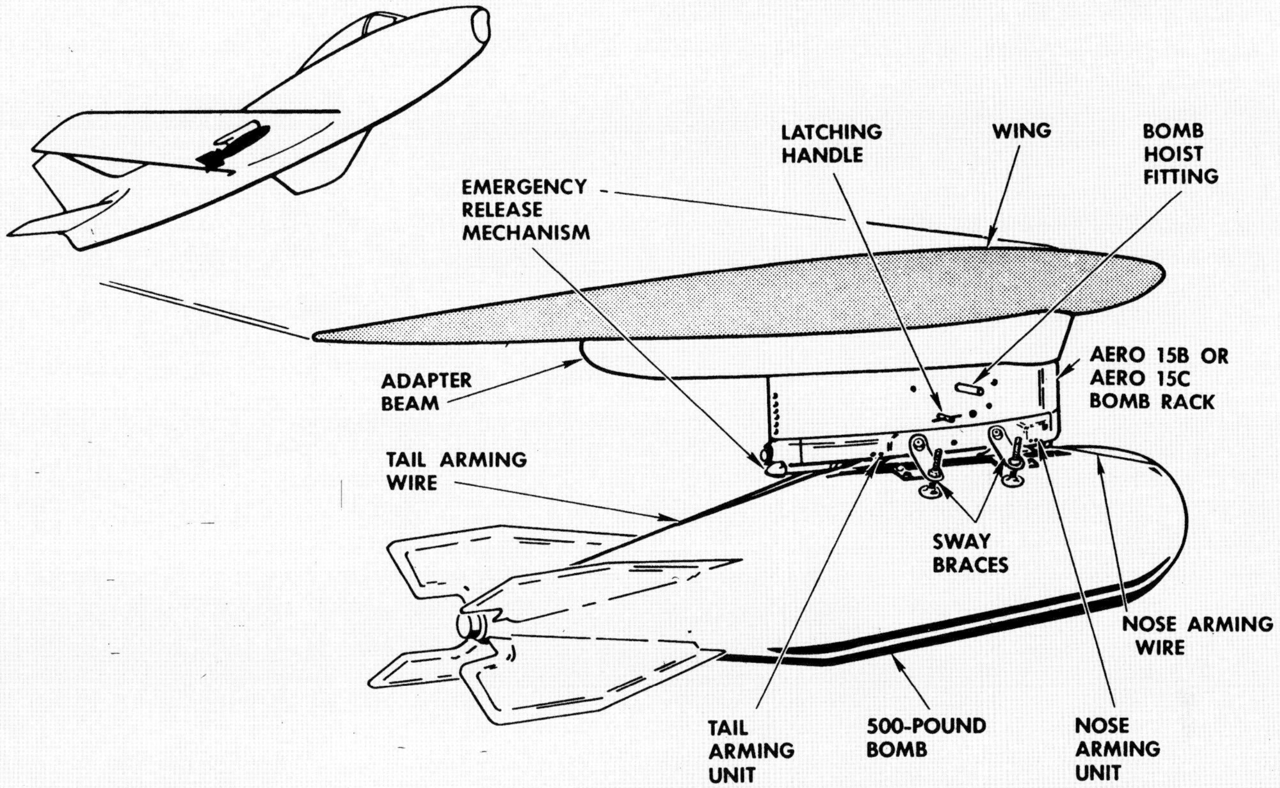

At left, 1,000 lb and 250 lb bomb fitted to the left-hand wing. (National Archives)

Below, FJ-3M with two 1,000 lb and 500 lb bombs during carrier catapult tests. (National Archives)

SIDEWINDER INSTALLATION

The first fleet aircraft to deploy the Sidewinder missile operationally on aircraft carriers was the FJ-3M Fury. It could only carry two Sidewinders, one on each outboard pylon, on specialized missile racks. The FJ-3M used an outboard adapter beam first followed by an Aero 3A guided missile launcher pylon and an Aero 3A guided missile launcher.

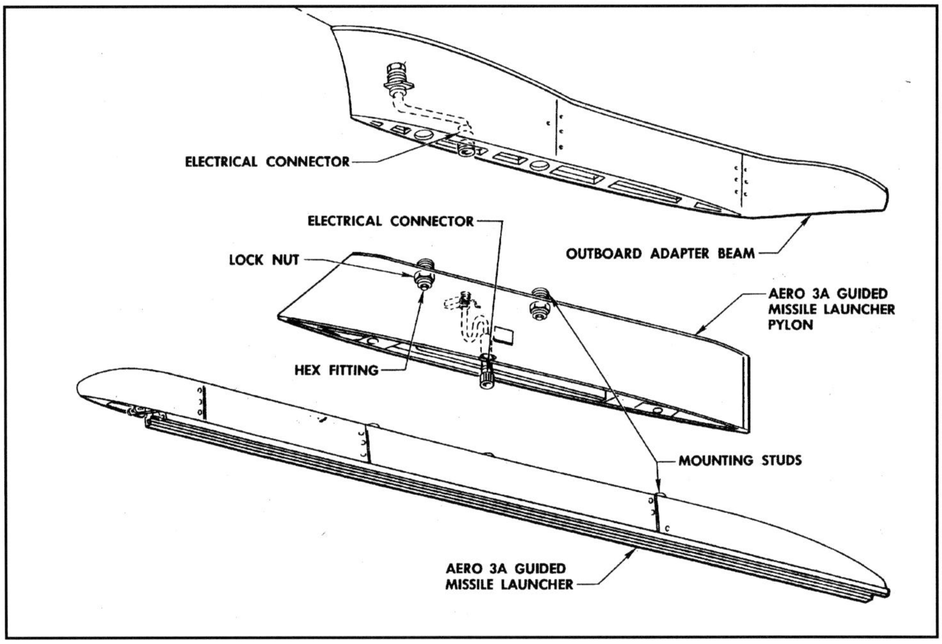

Below, flight test installation of the Sidewinder on the outer wing pylon of the FJ-3M Fury. (USN)

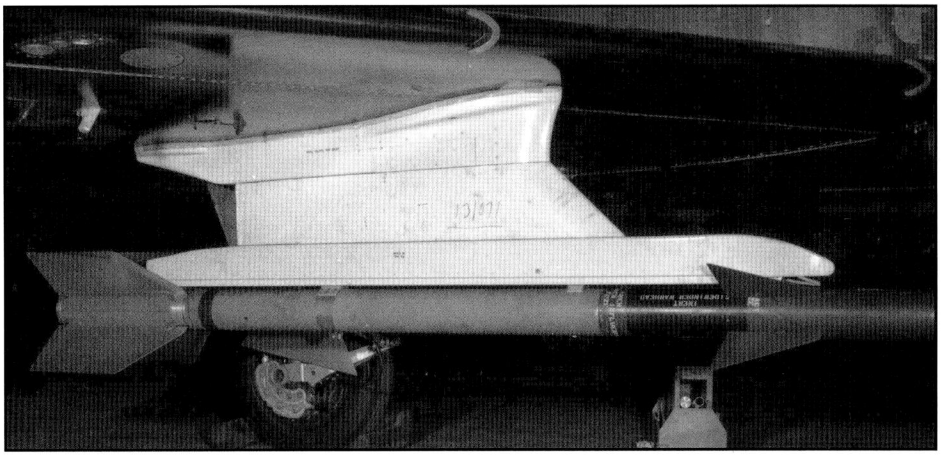

FJ-3D REGULUS I DRONE CONTROL AIRCRAFT

The primary purpose of the FJ-3D was its use as a drone control aircraft for the Regulus I missile. The equipment used in controlling the missile was the Radio Control System, AN/ARW-55, along with various controls and panels necessary to operate the equipment. Installation of this additional equipment in the airplane necessitated removal of some components and relocation of others. The guns were aimed by the fixed sight as the radar set, AN/APG-30, and the aircraft fire control system, Mk. 16 Mod 2, had been removed. The FJ-3D-2 Regulus II drone control aircraft was similarly configured.

The following table lists the major items of FJ-3 equipment which were not installed on the FJ-3D aircraft:

PART NUMBER	PART NAME
Mark 35 Mod 2	Control Panel
AT-233/APG-30	Antenna
PP-493/APG-30	Power Supply-Range Computer
MT-739/U	Mount
RT-181/APG-30	Frequency Converter-Transmitter
MT-707/APG-30	Mount
Mark 86 Mod 0	Computer
Mark 2 Mod 0	Ballistic Unit
CN-112/APG-30	Voltage Regulator
MT-708/APG-30	Mount
Mark 87 Mod 0	Range Rate Computer
814270E	Mount
Mark 25 Mod 0	Range Unit
1221044-A	Mount
C-775/APG-30	Control Panel
181-74222	Range Meter-Target Indicator Light
Mark 75 Mod 0	Power Supply
	Wave Guide and Associated Parts
	Crystal Current Meter
	Range Unit Volume Control
Mark 12 Mod 0	Left-hand Outboard Gun
Mark 7 Mod 0	Gun Feeder
34502-2	Ammo Booster
194-61285-10	Gun Muzzle Stabilizer
181-61029-21	Link Ejection Chute
181-61277-1	Case Ejection Chute
194-61025-1	Feeder Chute
194-61110-1	Ammo Box
73-202G-1	Gun Mount (Forward)
74-1220	Gun Mount (Rear)
	Associated Lines, Wiring, Hardware and Fittings

Modification from the FJ-3 to the FJ-3D neccessitated relocation and modification of certain components which are listed in the following table:

COMPONENT OR SYSTEM	DISPOSITION
Pneumatic System	Modified
Left-hand Gun Bay	Modified
AN/APX-6B	Moved to Left-hand Ammo Bay
AN/APA-89	Moved to Left-hand Ammo Bay
Left-hand Console	Modified
Pilot's Instrument Panel	Modified
Speed Brake Indicator	Relocated
Cockpit Pedestal	Modified
Associated Lines, Wiring, Hardware and Fittings	As Required

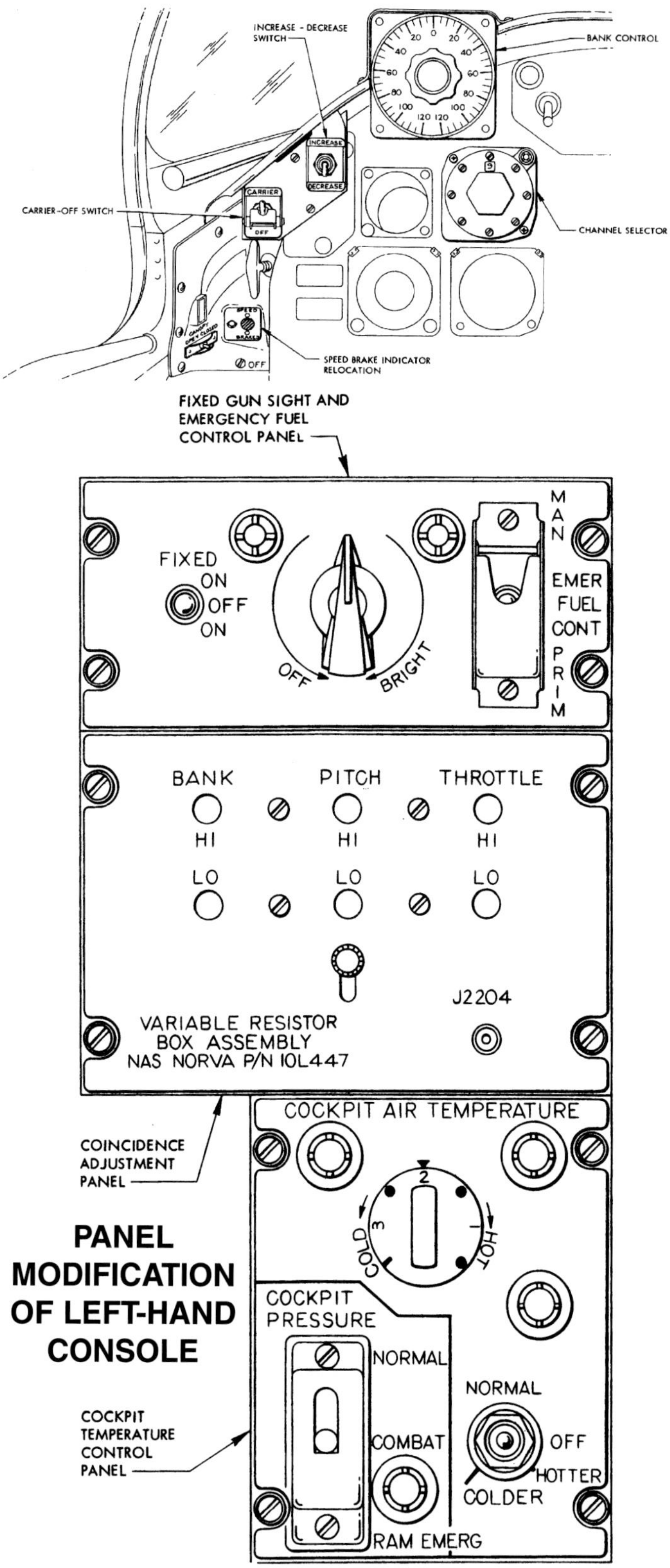

PANEL MODIFICATION OF LEFT-HAND CONSOLE

The special radio control equipment installed on the FJ-3D airplane is listed in the following table:

PART NUMBER	PART NAME
10D467	Ballast Weight
AN/ARW-55	Radio Control Equipment consisting of:
KY-111/ARW	Audio Coder
MT-1408/ARW	Mount
T-309/ARW-55	Transmitter
MT-822/ARC-27	Mount
AT-335/ARW	Antenna
C-902/ARW-55	Transmitter Control
MT-949/ARW-55	Mount
10D445	Pitch Control
10D433	Bank Control Assembly
10D464	Carrier-off Switch
10D464	Increase-Decrease Switch
10D446	Resistor and Condenser Assembly
10L447	Variable Resistor Assembly (Coincidence Adjustment Panel)
10D454	Gun Sight and Fuel Switch Panel
10D435	Channel Selector Assembly
10H536	Drone Throttle Control
	Associated Lines, Wiring, Hardware and Fittings

FJ-3D RADIO CONTROL ANTENNA AT-335/ARW-55

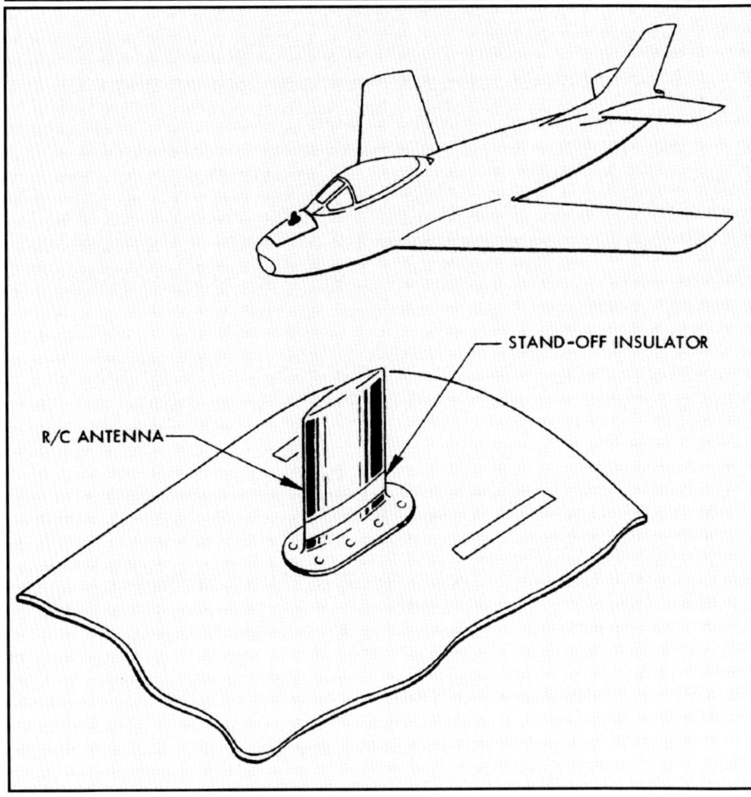

FJ-3D DRONE CONTROL AIRCRAFT INSTRUMENT PANEL

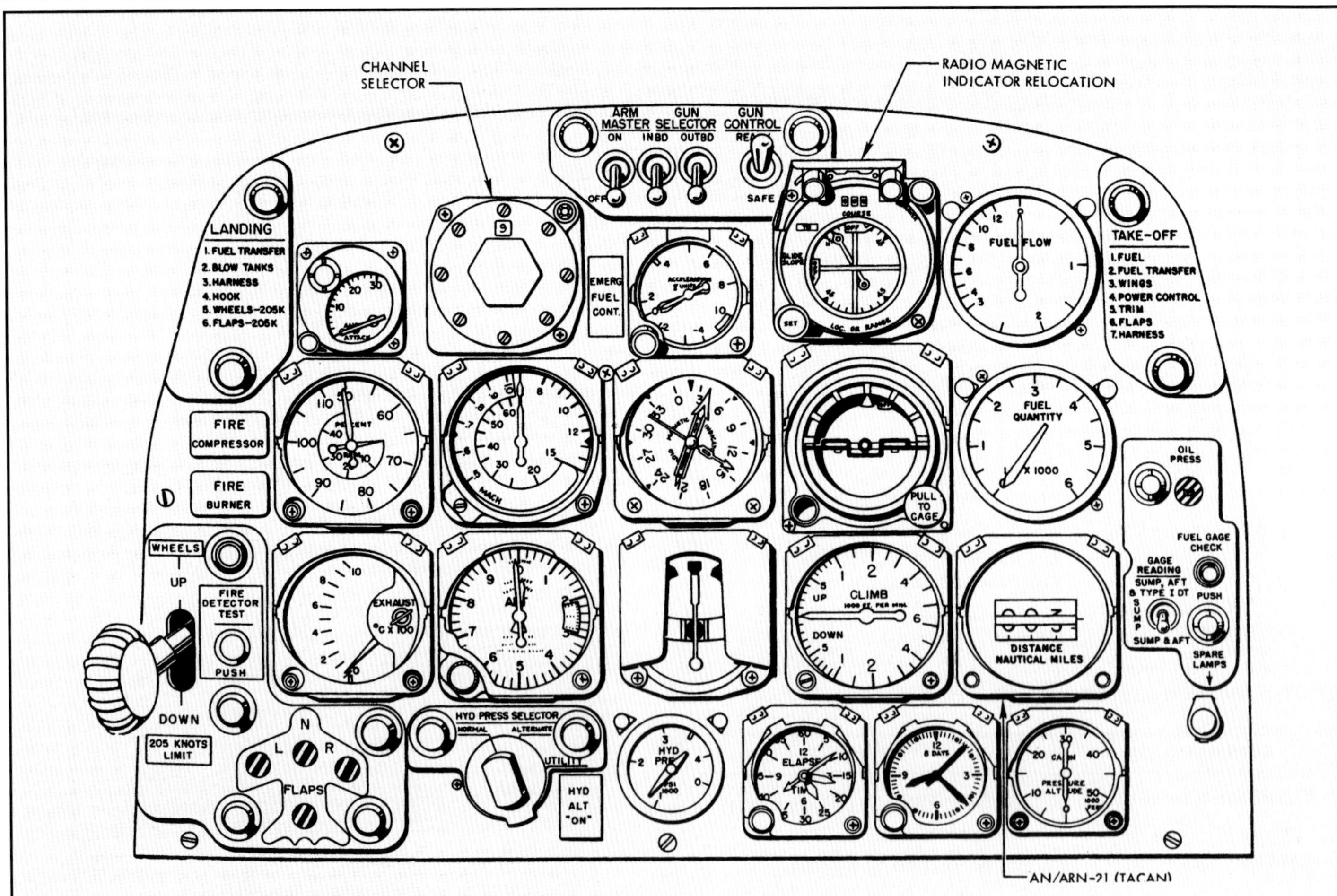

RADIO CONTROL GEAR IN FORWARD EQUIPMENT COMPARTMENT

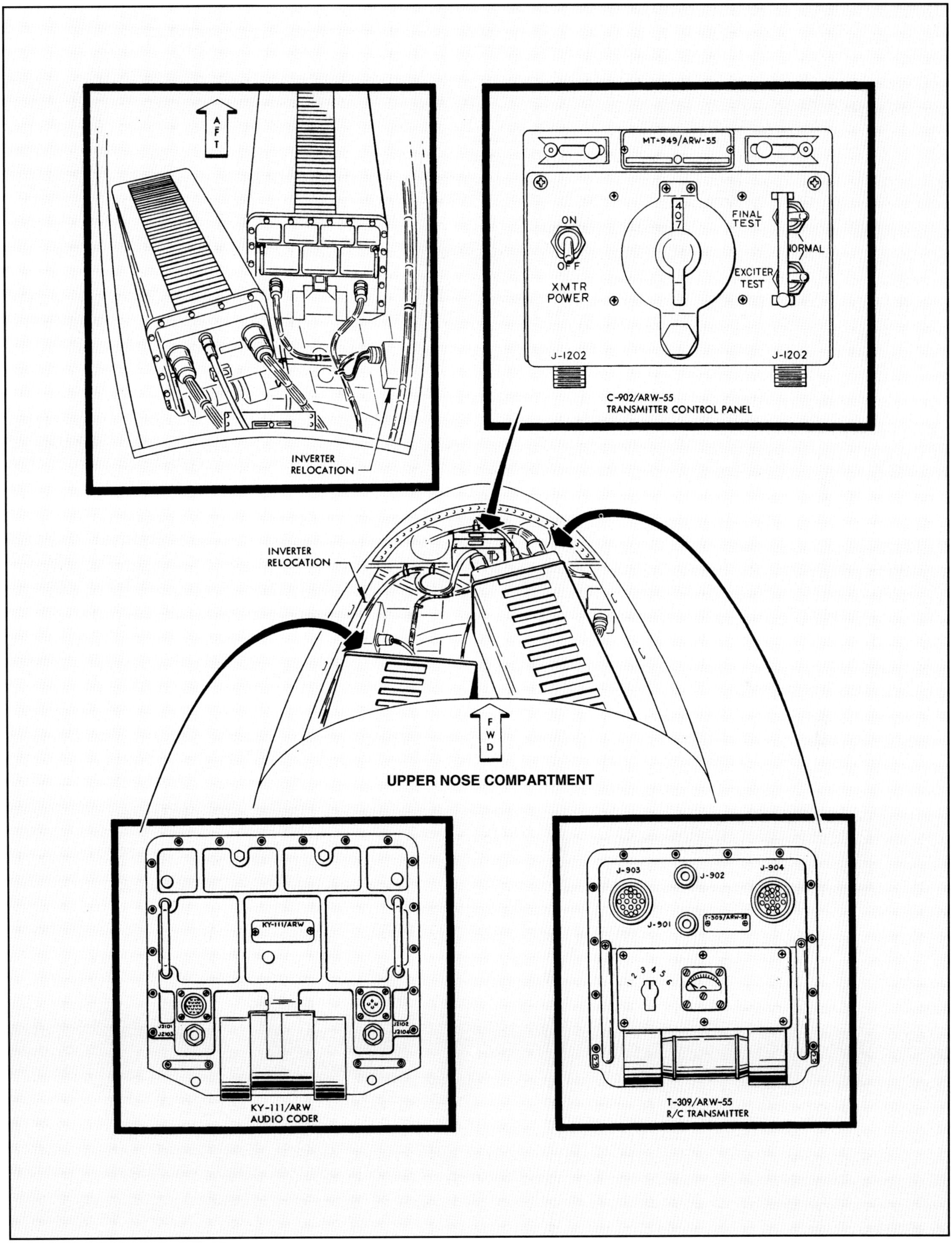

FJ-3D BALLAST INSTALLATION, LEFT - HAND GUN BAY

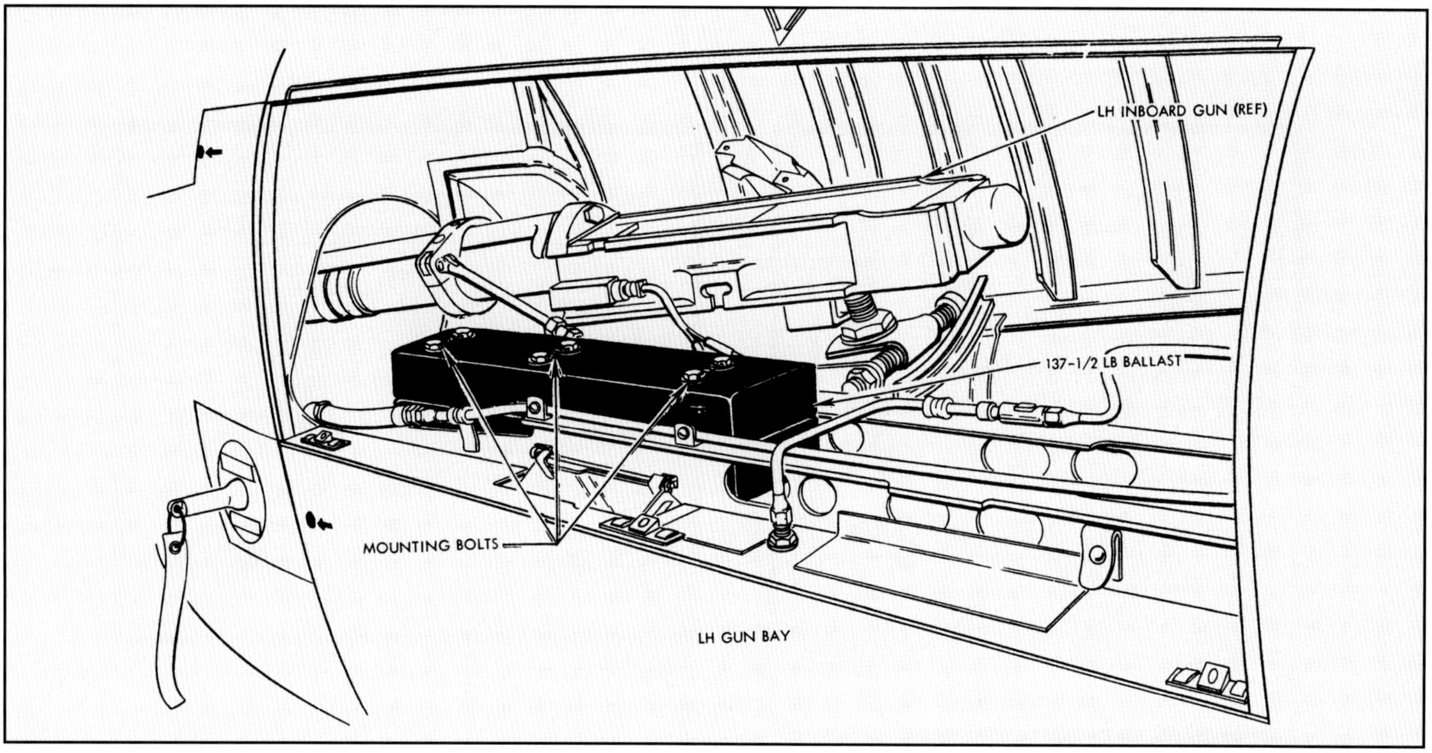

FJ-3D BALLAST INSTALLATION, LEFT - HAND GUN BAY

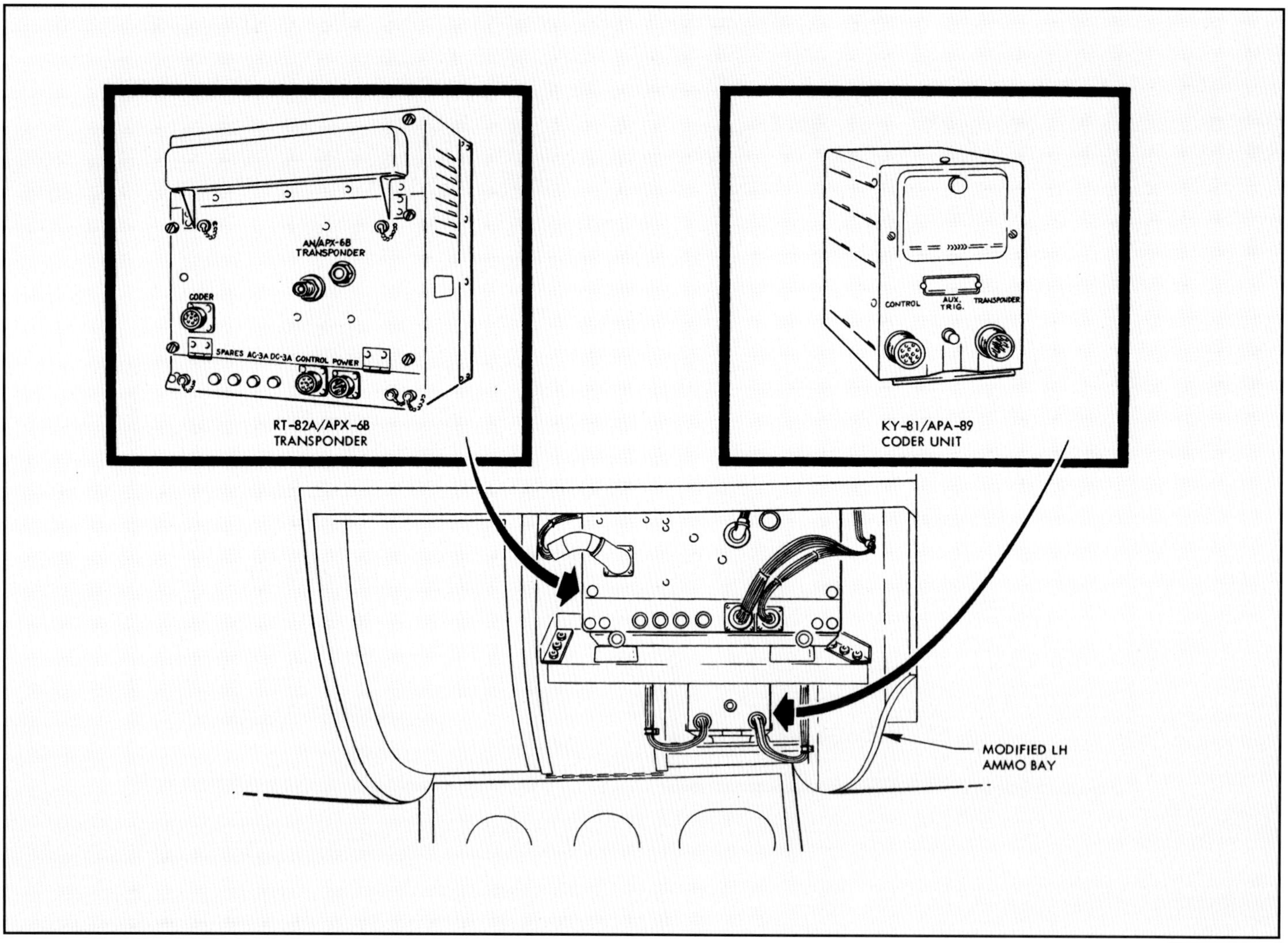

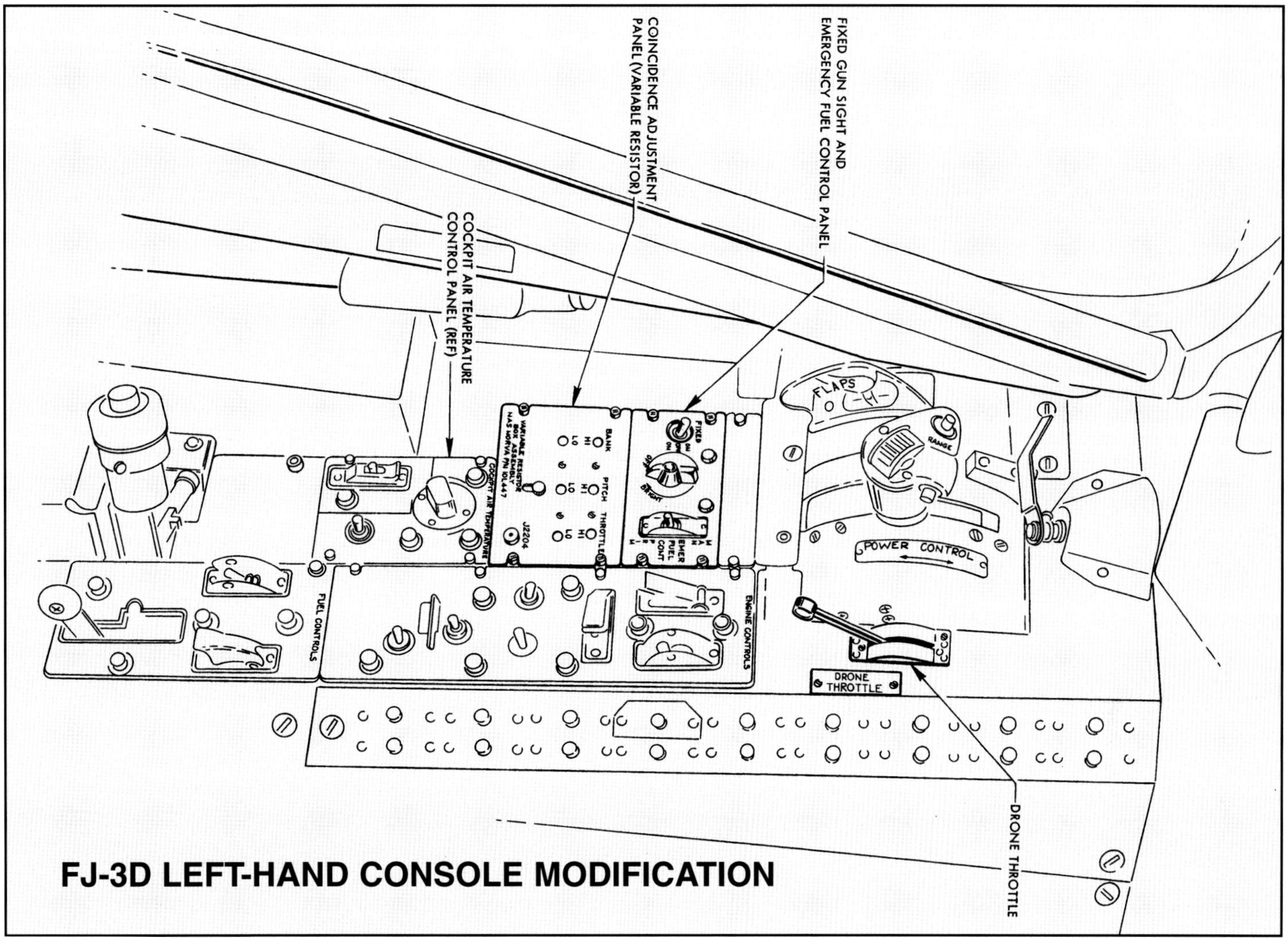

FJ-3D LEFT-HAND CONSOLE MODIFICATION

TRANSMITTER CONTROL C-902/ARW-55

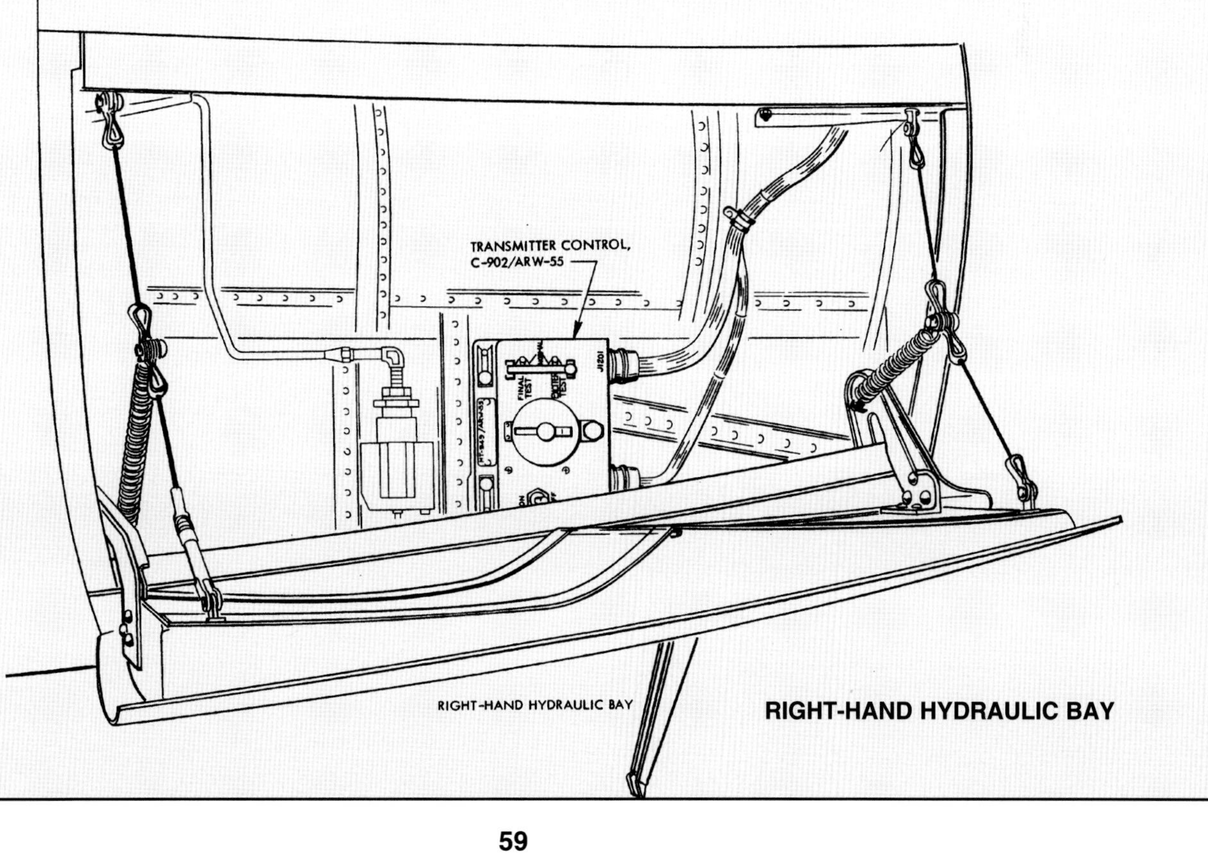

RIGHT-HAND HYDRAULIC BAY

NAVAL AIR TEST CENTER, NATC NAS PATUXENT RIVER, MD

The primary mission of the Naval Air Test Center (NATC) is to determine an aircraft's suitability for usage in the fleet. During the late '40s and early '50s, NATC project pilots and engineers were divided among four test divisions. Flight Test (FT) concentrated on aircraft and engine performance, stability and control and carrier suitability. Service Test (ST) evaluated operational and tactical suitability, emphasizing maintenance. Electronic Test (ET) was responsible for all the avionics equipment. Armament Test (AT) evaluated the aircraft as a weapons platform.

Additionally, the NATC supplied an evaluation team to the contractor's plant for a Preliminary Evaluation (NPE), a formal series of tests with the prototype models prior to the aircraft's arrival at NATC, and conducted fleet suitability evaluations using the test center's aircraft. Finally, the Service Test Division (ST) aircraft and personnel were later involved in the Fleet Introductory Program (FIP), which trained the first squadron, its pilots, and ground crews scheduled to operate the new aircraft.

Bottom, four FJ-3 Furies (BuNos 135786, 787, 788, and 790) assigned to NATC and used during the FIP program. (USN)

NATC

Above, NATC FJ-3 BuNo 135778 while assigned to Armament Test Division (AT 22). Intake lip, gun troughs and landing gear were natural metal. Landing gear was painted silver. (via Lionel Paul) At right, NATC FJ-3 BuNo 135776 from Flight Test Division (FT) launches from the USS Coral Sea (CVA-43) during carrier suitability tests. (USN) Below, NATC FJ-3 BuNo 135788 while assigned to Service Test Division (ST) in January 1955 aboard Coral Sea during carrier suitability tests. (National Archives)

NATC

Above and below, an FJ-3M and an FJ-4 Fury during in-flight refueling tests with a NATC AJ-2 Savage. (National Archives)

At left, Naval Air Test Center Flight Test (FT) FJ-3 Fury NATC/143 in flight. (National Archives)

FLEET AIR GUNNERY UNIT, FAGU

The Fleet Air Gunnery Unit (FAGU) was established at NAF El Centro, CA, in the summer of 1952. FAGU utilized the FJ-3 Fury as a gunnery training aircraft for jet fighter and attack pilots.

The unit's mission was to train experienced pilots in standardized methods and advanced techniques in air gunnery, bombing and rocketry. These pilots would return to their units and pass on their knowledge to the other pilots. A typical class consisted of five pilots and lasted four weeks. The course consisted of 175 hours of ground school and 45 hours of gunnery and rocketry work. Pilots practiced gunnery and rocket fire at ground targets, air-to-air gunnery, field carrier landings, night flying, ground control intercept missions and air support hops.

Below, early production FJ-3 assigned to FAGU had a red and white checkered tail, white nose stripes and a red lightning bolt bordered in white. (USN) Bottom, FAGU FJ-3 Fury BuNo 141408 taxi's at NAF Fallon, NV. Nose stripes were white and black. Vertical tail was covered in a red and white checkerboard. Nose lightning bolt was red bordered by white. (Charles V. Cable via Tailhook)

NAVAL AIR MISSILE TEST CENTER (NAMTC) / NAVAL MISSILE CENTER (NMC)

Although the Regulus came to the Naval Air Missile Test Center (NAMTC) in 1947 in the form of a model drop test program, it was not until two years later that an actual flight test program began. The XSSM-N-8 Regulus I was a 500-mile surface-to-surface, subsonic, guided (cruise) missile.

Initial flight testing was conducted at Muroc Dry Lake (Edwards AFB) with zero-length launcher tests first performed at NAMTC. By March 1953, the Regulus flight test program at Point Mugu and Edwards had completed forty flights. Results showed that an early Fleet deployment was possible. The CNO then authorized the Regulus Assault Missile (RAM) program. A selected group of cruisers and carriers were picked to field the weapon. These were the:

USS Los Angeles (CA-135), USS Helena (CA-75), USS Toledo (CA-133), USS Macon (CA-132), USS Princeton (CVA-37), USS Hancock (CVA-19), USS F.D. Roosevelt (CVA-42), USS Saratoga (CVA-60), USS Lexington (CVA-16), USS Randolph (CVA-15), USS Shangri-La (CVA-38), and USS Ticonderoga (CVA-14)

The RAMs would be guided via radio control by specially modified F9F-8 Cougars and FJ-3D Furies.

From 18 through 30 June 1953, NAMTC conducted submarine launch tests from the USS Tunny (SSG-282). Eventually, five submarines were converted as Regulus delivery systems called SLAM (Submarine Launched Assault Missile). These were the:

USS Barbero (SSG-317), USS Tunny (SSG-282), USS Growler (SSG-577), USS Grayback (SSG-574), and USS Halibut (SSGN-587)

Regulus I testing continued into 1956 when the Regulus II flight test program began on 19 March 1956. The Mach 1.5 Regulus II would be controlled by the FJ-3D2 aircraft. Again, the first flights were conducted at Edwards AFB. The first launch from Point Mugu occurred on 5 June 1958.The first Regulus II submarine launch occurred on 16 September from the USS Grayback. The Regulus II program was cancelled shortly thereafter.

Even though the Regulus program was terminated, the drone control FJ-3s continued to be used at Point Mugu through 1962. NAMTC was redesignated the Naval Missile Center (NMC) on 7 January 1959.

Below, a Regulus I test launch from the Naval Air Missile Test Center, NAS Point Mugu, CA, in 1957. (USN)

Above, FJ-3D BuNo 136112 at Point Mugu on 18 May 1962. (Doug Olson) Below, Regulus II test launch at Edwards AFB. (via Tommy Thomason) Bottom, Pacific Missile Range FJ-3D BuNo 135995 at Point Mugu on 22 April 1961. Colors were engine blue-grey fuselage, yellow wings with a wide da-glo red stripe, da-glo red fin, yellow horizontal stabilizer and yellow drop tanks and refueling probe. (William Swisher)

NAVAL ORDNANCE TEST STATION (NOTS) / NAVAL WEAPONS CENTER (NWC)

Below, NWC China Lake FJ-3D BuNo 136064 in November 1961. (Clay Jansson) Bottom, FJ-3D BuNo 135788 over the Sierras. (R.W. Harrison)

Naval Ordnance Test Station (NOTS) was established to conduct rocket and aviation ordnance testing and development under the leadership of CALTECH. The Sidewinder was arguably the most significant success produced through this union. Although Sidewinders were first deployed on FJ-3M Furies, FJ-3/3Ms were not used at China Lake for any significant testing. However, the Naval Weapons Center at China Lake (NWC) did operate the FJ-3D drone control aircraft..

GUIDED MISSILE GROUP ONE, GMGRU-1

GMGRU-1 was established on 16 September 1955 when VU-3 RAM DET and GMU-52 were combined. The unit, under the command of CDR Millard, was originally stationed at NAS North Island, CA, but most of its early work was conducted in the Mojave Desert from MCAAS Mojave and later San Nicolas Island. The main mission of the unit was the control and delivery of Regulus ship and submarine launched missiles. The F9F-2KD and F9F-5KD drones were used primarily to train drone control pilots which flew TV-2Ds, F9F-6Ds and FJ-3Ds.

GMGRU-1 first received the FJ-3D in May 1956. Senior pilots conducted familiarization flights at NAS Moffett Field-based VC-3 and then became the group's instructors. In addition a ground school of such was conducted by VF-121 at NAS Miramar. During this transition period one Fury stalled on take-off and caught fire after the pilot escaped with minor burns.

The unit, along with Submarine Division 51 from Port Hueneme, CA, moved to Hawaii on 20 July 1956. The unit was based at Barbers Point, Hawaii, but also operated from ALF Bonham at Barking Sands, Kuai. The Regulus test launches and recoveries were also conducted from ALF Bonham. The first Regulus was launched on 16 October 1956, followed by eleven more training launches.

GMGRU-1 RAM Det Alpha (Regulus Assault Missile) deployed

Below, four GMGRU-1 FJ-3Ds, BuNos 135779 (ZZ/9), 136095 (ZZ/1), 136097 (ZZ/7) and 136107 (ZZ/8) in flight over the Pacific. Blue aircraft had an engine blue-grey fuselage, yellow wings with a wide da-glo red stripe, da-glo red fin, yellow horizontal stabilizer and yellow drop tanks and refueling probe. (C.M. Hansen via Lawson)

Above, GMGRU-1 FJ-3Ds BuNos 135950 (ZZ/4), 136107 (ZZ/8), and (ZZ/6) in flight. (C.M. Hansen via Lawson) Below, GMGRU-1 Det CONUS FJ-3Ds BuNos 136018 and 135995 at NAS Point Mugu. Wing tanks were faded da-glo red. (William Swisher) Bottom, three GMGRU-1 FJ-3Ds BuNos 135779 (ZZ/9), 136097 (ZZ/7) and 136107 (ZZ/8). (C.M. Hansen)

with the FJ-3D for the first time on 13 November 1956 aboard the USS Shangri-La (CVA-38). Two pilots, LTJGs Duane H. Myrin and Bill Allen, and fifteen enlisted personnel along with two FJ-3Ds were assigned spaces with VF-24. The squadron's first mission from CVA-38 was the guidance of a Regulus launched from the USS Toledo (CA-133).

GMGRU-1 RAM Det Golf deployed aboard the USS Lexington

(CVA-16) on 14 April 1957. Assigned to VF-121, the Det included two aircraft, fifteen enlisted men and three pilots: LCDR V.F. Foresberg, LT R.P. Blount and LTJG A.J. Thayer. On 17 June a Regulus tactical missile was launched from the USS Helena (CA-75) and LT Blount took over radio control of the missile for its 180 nm run to target.

In 1957, GMGRU-1 Det Sierra was established at NAF Naha, Okinawa, for WestPac deployed ships. The Det was commanded by LT M.E. Smith.

GMGRU-1 RAM Det Mike deployed aboard the USS Ticonderoga (CVA-14) in October 1957. Assigned to VF-91, the Det was comprised of two aircraft, fourteen enlisted men and LTs Roy Kraft, Jim Rice and LTJG Dave Newbro. Once

Above, Regulus I fired from the USS Helena (CA-75) on 17 June 1957 and controlled by Det Golf FJ-3Ds. (USN) Below, GMGRU-1 Det Golf FJ-3D BuNo 136005 aboard CVA-16 in 1957 (USN) Bottom, GMGRU-1 FJ-3D BuNo 135986 overflies the Lexington in 1957. (USN)

GMGRU-1 FJ-3D gear-up grass landing at NAF Naha, Okinawa, in 1958. (USN/Philip Sisney via Tailhook) Below, GMGRU-1 FJ-3D BuNo 135884 during carrier operations. (Joseph A. Gryson via Tailhook)

on station the Det was offloaded and operated from Naha for the remainder of the cruise.

On 8 March 1958, GMGRU-1 RAM Det Charlie deployed aboard the USS Shangri-La (CVA-38) with two FJ-3Ds, LTs D.L. Whitman, D.H. Edgren, F.J. Orrik and twenty enlisted men. Regulus operations were conducted off Okinawa for a major SEATO exercise. The Det controlled a launch from the Tunny (SSG-282) on 22 May, and from the USS Toledo (CA-133). After the exercise the Det was shore based at NAHA.

The final FJ-3D deployment was Det Lima aboard the USS Lexington (CVA-16) on 17 July 1958. Again two FJ-3Ds were deployed, along with LT Phil Canup and LTJG Todd Vieregg.

Above, Regulus I being controlled by a TV-2 with a FJ-3D armed chase aircraft in case the test goes awry. (USN) Bottom, two GMGRU-1 FJ-3D BuNos 136010 (DA/5) and 136100 escort a NOLO KD-9 off Oahu on 13 May 1957. (Bob Lawson)

In June 1956 GMGRU-1 estab-

lished a CONUS (continental United States) Det at NAS Point Mugu, CA, commanded by LCDR Len Plog. The Det provided aircraft support for West Coast submarine and cruiser training and R&D support for Regulus, Terrier and Sparrow III development. An FJ-3D and LT H.K. Chamberlain were lost on 16 October 1957 during a night takeoff from Point Mugu. In 1958, the FJ-4B replaced the Det's FJ-3Ds.

On 1 September 1960, GMGRU-1 was disestablished and its assets were received by Utility Squadron One (VU-1).

Above, GMGRU-1 CONUS Det FJ-3D BuNo 135962 at the NAMTC, NAS Point Mugu, CA, in December 1956. Trim stripes were powder blue and gold. (USN) Below, two GMGRU-1 FJ-3Ds BuNos 136100 (DA/9) and 136010 (DA/5) pass by the USS Shangri-La (CVA-38) on 13 May 1957. (Bob Lawson)

UTILITY SQUADRON ONE, VU-1 "UNIQUE ANTIQUERS"

VU-1 was originally established as VJ-1 on 5 October 1925 at NAS San Diego, CA. The squadron fulfilled the missions of aerial photography, simulated attacks, target towing and aerial and anti-aircraft gunnery training. The squadron also participated in the Alaskan Aerial Surveys of 1926 and 1934.

In September 1939, a VJ-1 detachment moved to Ford Island, Pearl Harbor, T.H. By June 1940 Ford Island was home for the entire squadron. After the attack on Pearl Harbor, VJ-1 provided utility services throughout the Pacific.

On 31 July 1944, the squadron moved to NAS Moffett Field, CA, where it was redesignated Utility Squadron One (VU-1) on 15 November 1946. VU-1 was disestablished on 30 April 1949.

VU-1 was re-established at NAS Barbers Point, T. H., on 20 July 1951, from the VU-7 detachment already stationed there. The squadron was originally equipped with five JD-1s (A-26s), five TBM-3Us, and one SNB-2P. The F9F Panther arrived in 1953 and the Cougar in 1956. When GMGRU-1 was disestablished on 1 September 1960 its assets, including the FJ-3Ds were transferred to Utility Squadron One.

The squadron provided fleet utility services until redesignated Fleet Composite Squadron One (VC-1) on 1 July 1965.

Below, three VU-1 FJ-3Ds (BuNos 135784 UA/6, 136107 UA/8 and 136087 UA/7) along with a VU-1 FJ-4B in flight off Hawaii. (C.M. Hansen via Lawson)

Above, VU-1 FJ-3D BuNo 136095 from NAS Barbers Point flying wing with a Cougar in 1961. (C.M. Hansen via Bob Lawson) Below, VU-1 FJ-3Ds in storage at NAS Alameda, CA, in April 1967; BuNo 135884 (left), 15986 (right). (Jim Farmer) Bottom, FJ-3D/DF-1C BuNo 135995 at Alameda on 10 July 1965. (Clay Jansson)

GUIDED MISSILE GROUP TWO, GMGRU-2/GMSRON-2

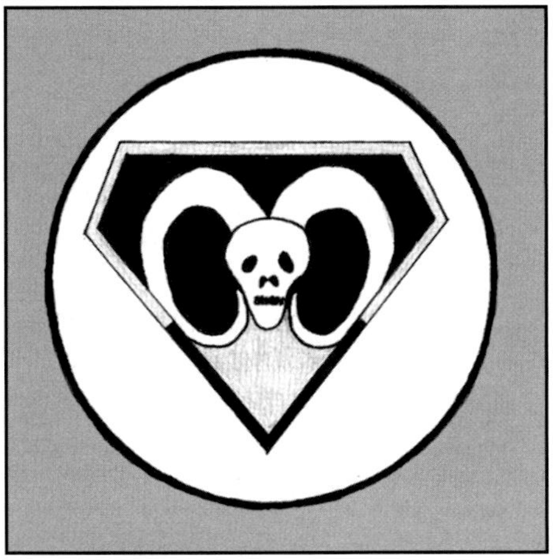

On 23 September 1955, VU-4 RAM DET and GMU-53 combined to form GMGRU-2 at Chincoteague, VA.

In 1956, Atlantic Fleet Regulus I launches and RAM training were extensive. Launches were conducted from the USS Macon (CA-132), USS Randolph (CVA-15) and USS Barbero (SSG-317). Macon launched five Regulus I missiles which allowed the squadron to conduct RAM training after the operational portion of the Regulus flight was completed.

Randolph flight operations commenced next. The squadron deployed RAM Det 36 with four pilots (LT A.J. Monger, LTJGs A. O'Brien and S. Olmstead, ENS R. Pekkanen), three aircraft and fourteen enlisted men. A training missile launch was performed on 15 June 1956 and a tactical missile launch on 18 June. The Det and Randolph set sail for the Eastern Mediterranean on 14 July 1956.

On the trip across, a bad catapult shot claimed the life of ENS Pekkanen. He was replaced by LTJG Dave Leue. Once in the Med the squadron participated in air power demonstrations and in November and December monitored the Suez Crisis. Det 36 returned to Chincoteaque on 18 February 1957.

A short deployment was then conducted in late February for Operation Snowball aboard the USS F.D. Roosevelt (CVA-42) with one Regulus I, three pilots (LTs L. Dion, W. O'Brien, W.D. Laurentis) and three FJ-3Ds. The purpose of the cruise was to test cold weather operations of the system.

In May 1957, a Det was sent to NAF Mayport, FL, to support three missile launches from the USS Saratoga (CVA-60). Two familiarization launches were conducted prior to a Sidewinder intercept launch. The three Charlie aircraft FJ-3Ds launched with two Sidewinder-armed F4D-1 Skyrays who splashed the Regulus after expending all eight Sidewinders.

GMGRU-2 was redesignated Guided Missile Service Squadron Two (GMSRON-2) on 1 July 1958. Since carrier RAM was no longer utilized, missile launching capability was handed over to GMU-51. GMSRON-2 continued to provide missile recovery and drone services.

Below, GMGRU-2 FJ-3Ds and TV-2D successfuly bring a Regulus I back to Chincoteague. (USN)

Above, two TV-2Ds and two FJ-3Ds from GMGRU-2 escort a Regulus I across the runway at NOTS NAAS Chincoteague on 17 April 1957 during a demonstration for defense officials. (USN) At right, Regulus I tactical missile launch off the North Carolina Coast from the heavy cruiser USS Macon (CA-132). The non-recoverable tactical missiles were painted dark blue, while training missiles were red with white wings. (Vought) Bottom, 1956 GMGRU-2 squadron photo at NAAS Chincoteague. The XO, CDR John "Mickey" Braun, is second from left and the CO, CDR W.B. "Bill" Coley, is third from the right. (USN)

UTILITY SQUADRON TWO, VU-2

Utility Squadron Two (VU-2) was initially a detachment of Utility Squadron Four (VU-4) at NAS Quonset Point, RI. As the demand for utility services increased, the VU-4 Det was established as VU-2 on 8 January 1952. The squadron was originally equipped with the JD-1 (A-26) Invader and the F9F Panther. The F9F Cougar followed shortly thereafter. FJ-3 Furys were also added and the unit moved to NAS Oceana, VA, in June 1960. Detachment Alfa remained at NAS Quonset Point with the JDs. In 1961, the supersonic F8U-1 Crusader replaced the squadron's Furys.

The squadron provided fleet utility services until redesignated Fleet Composite Squadron Two (VC-2) on 1 July 1965.

Top to bottom, two photos of VU-2 FJ-3 BuNo 135872 which had a small red fin stripe. (Lionel Paul) BuNo 136144 with blue trim. (Ron Picciani) BuNo 136137 on 18 March 1963. (Doug Olson)

UTILITY SQUADRON THREE, VU-3

The second Utility Squadron Three (VU-3) was established on 20 December 1948 at NAS Santa Ana, CA. It was assigned the mission of furnishing radio-controlled drone aircraft for fleet target practice. The squadron moved to NAS Miramar in 1950, NAAS Ream Field in 1951 and NAAS Brown Field in 1956. The FJ-3D2 was acquired to control the KDA jet targets and the squadron moved to NAS North Island in 1961 where the F4D-1 replaced the Furies.

During 1954-1955 VU-3 operated a Regulus I RAM detachment using F9F-6D control aircraft. The FJ-3D/D2 was not used by the RAM det.

Below, VU-3 flightline at Brown Field on 30 April 1960. FJ-3D2 BuNos 135867 (UF/14), 136025, 135783, and 135809 (UF/19). (Clay Jansson) Bottom, FJ-3D2 BuNo 135809 on 30 April 1960. Fuselage was engine blue-grey. Wings, tanks, horizontal tail, and aft fuselage body was yellow. Vertical tail and wide wing stripe was da-glo red. (William Swisher)

FLEET COMPOSITE SQUADRON THREE, VC-3

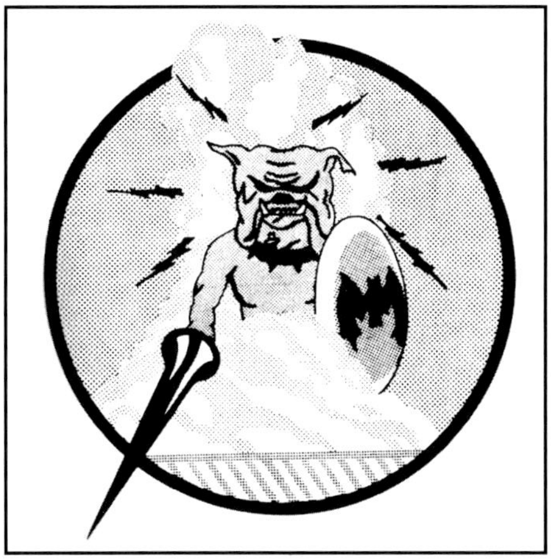

On 2 May 1949, Det 1 Fleet All-Weather Training Unit Pacific (FAWTUPAC) was morphed into Composite Squadron Three (VC-3). Under the command of LCDR Lewis R. Hardy, the squadron was established at North Island, NAS San Diego, CA, with five F6F-5Ns, six TBM-3E/3Ns, an AD-2Q and a SNB-5. VC-3 moved to its new home in LTA Hangar One, NAS Moffett Field, CA, in October 1949.

VC-3 was responsible for providing night fighter and night attack detachments to Pacific Fleet carriers. In May 1950, the attack division was notified of its transfer to the newly-established VC-33 based in San Diego. By the end of 1952, the squadron was operating six types of fighter aircraft: the F6F-5N, F4U-4,

Above, early slatted-wing VC-3 FJ-3 during maintenance in a LTA hangar at NAS Moffett Field, CA, in January 1955. (National Archives) Below, CDR R.W. "Duke" Windsor, XO of VC-3, leads four FJ-3s in flight from NAS Moffett Field, CA, in January 1955. (National Archives)

Above, VC-3 FJ-3 BuNo 135810 was trimmed in white with natural metal leading edges and wing tank tips. Note extended tailhook. (William T. Larkins) Below, VC-3 FJ-3 NP/80 undergoing an engine check in January 1955. (National Archives) Bottom, VC-3 FJ-3 BuNo 135832 in 1955. (William T. Larkins)

F4U-5N, F3D-1, F2H-2B, and the F2H-3. During the Korean War, night fighter detachments of F4U-5N Corsairs were dispatched to the war zone and LT G.P. Bordelon became a nocturnal ace in the type with five kills.

On 1 June 1954, CDR J.D. "Jig Dog" Ramage assumed command of VC-3 and with him an additional squadron mission. Named "Project Checkout". the squadron assumed the mission of training and qualifying Pacific Fleet pilots in current and newly arriving fighter jet aircraft. No more kick-the-tires-and-light-the-fire familiarization. By Christmas 1954, VC-3 had six F7U-3 Cutlasses, six FJ-3 Furies and six F9F-6 Cougars for jet transition training. In the spring of 1956, the F4D-1 Skyray arrived and the squadron was redesignated Fleet All-Weather Fighter Squadron Three (VF(AW)-3) on 1 July. F3H-2N Demons, A4D-1 Skyhawks and F8U-1 Crusaders were added prior to the squadron's disestablishment on 2 May 1958.

AIR DEVELOPMENT SQUADRON THREE, VX-3

Air Development Squadron Three (VX-3) was established on 20 November 1948 by the merger of CVLG-1, squadrons VF-1L and VA-1L. The mission was to evaluate by operational testing new and existing carrier type aircraft and recommend tactical employment, training procedures, training aids, countermeasures and to assist other naval activities or agencies as directed.

The squadron operated virtually every type aircraft in the Navy inventory including the FJ-3 Fury. In late 1954, the squadron deployed their F9F Cougars and FJ-3 Furies aboard the USS Ticonderoga (CVA-11). They deployed again in June 1955 for carrier qualifications aboard the USS Lake Champlain (CVA-36) with F9Fs, F2Hs, FJ-3s, and F7Us. The winter of 1955 saw the squadron aboard the USS Tarawa (CVA-40) with a brief stay at Guantanamo Bay, Cuba. In 1956, the FJ-3s were upstaged by arrival of the Demon, Skyray, Tiger and Crusader.

The squadron operated out of NAS Atlantic City, NJ, until May 1958 when they moved to NAS Oceana, VA. The squadron was disestablished on 1 March 1960.

Above, unknown civilian dignitaries pose in front of a VX-3 FJ-3 BuNo 135796 XC/94. (USN via R.J. Donaldson/Tailhook) Below, two VX-3 FJ-3s, BuNos 125794 XC/92 and 135796 XC/94, in flight over the Atlantic. (USN via R.J. Donaldson/ Tailhook)

Above, VX-3 FJ-3 BuNo 135793 on 20 May 1955. (via Burger) Below, VX-3 FJ-3 BuNo 135795 in 1956. (William T. Larkins) Bottom, two VX-3 FJ-3s BuNo 135794 XC/92 and XC/96 escort a squadron Skyraider. (USN)

UTILITY SQUADRON FOUR, VU-4 "DRAGON FLYERS"

VU-4 was originally established as VJ-4 on 15 November 1940 at NAS Norfolk. Subsequent home bases were Guantanamo, San Juan, Squantum, Chincoteague and Oceana. The squadron had operated over thirty types of aircraft from its first PBY-1 Catalina to the supersonic F-8 Crusader.

The mission of VJ-4 during WWII was to fly anti-submarine patrol and conduct rescue operations. After the war the squadron was redesignated Utility Squadron Four (VU-4) and its mission was changed to that of fleet support in air-to-air and surface-to-air weapons training as well as training fleet air intercept controllers and radar and ECM operators.

In 1952, the squadron added F6F-5K and F9F-5K drones to its inventory for use as realistic targets for the fleet. On 5 June 1958, P2V-5 and UB-26J aircraft were obtained for launching jet powered KDA drones. The Delmar towed target arrived in February 1960, one month after the squadron became responsible for Replacement Pilot Training for FJ-3M Furys. This lasted until August 1961 and 65 pilots were transitioned.

UTRON FOUR was redesignated VC-4 on 1 July 1965.

Below, VU-4 FJ-3M BuNo 141440 in 1960 was used for replacement pilot training with da-glo nose, tail, outer wings & tanks. Fuselage stripe and rudder were blue. (via Burger)

UTILITY SQUADRON FIVE, VU-5 "WORKHORSE OF THE FLEET"

Fleet Utility Squadron Five (VU-5) was established on 16 August 1950 in order to provide utility services to the fleet in and around the islands of Japan. As the squadron and its services expanded, UTRON 5 was required to send detachments to carriers and bases as far from its home base of NAS Atsugi as NAS Cubi Point, Philippines. VU-5's primary mission was to provide targets for both the aerial and surface components of the fleet. Other missions consisted of photo services, adversary missions, airborne control of surface launched missiles, and carrier onboard delivery (COD) services.

The squadron operated a number of FJ-3D2 (DF-1D) drone control aircraft into the early 1960s prior to becoming Fleet Composite Squadron Five (VC-5) on 1 July 1965.

Above, VU-5 FJ-3D BuNo 135997 drone controller follows a F9F-5KD target on a taxiway at NAS Atsugi on 3 november 1961. (T. Matsuzaki) Below, 135997 on 19 May 1962 on Okinawa. (via Lionel Paul) Bottom, VU-5 DF-1D BuNo 135797 after retirement at NAS Alameda, CA, on 10 July 1965. (via Clay Jansson)

UTILITY SQUADRON EIGHT, VU-8 "RED TAILS"

Above, GMSRON-2 Regulus I on its launch platform. (National Archives)

GMSRON-2 (Guided Missile Squadron Two), was established on 1 July 1958 at NAS Roosevelt Roads, Puerto Rico. The Regulus missile continued on as a target alongside of F9F and KDA targets with control aircraft being FJ-3D Furies.

In 1959, Det Alpha was operated out of NAS Mayport, FL, using the Regulus as the target aircraft.

The F9F-6K would be launched automatically after the taxi pilot left the aircraft. Two piloted FJ-3D/DF-1D Furies would pick up the Cougar in flight and control it remotely. Once in the target area, the Furies departed and the drone flies a predetermined target pattern. If not destroyed by Sidewinders or Sparrows, the Cougar was re-aquired by the Furies and returned to base where a DT-28 would control the aircraft to a safe landing.

The KDA-1 were carried to the target area by either a P2V-5F or a JD-1D, then dropped and controlled by the Furies. If the KDA-1 survived the mission they would ditch in the sea and then be picked up by a squadron helicopter.

GMSRON-2 was redesignated Fleet Utility Squadron Eight (VU-8) on 1 July 1960 and was redesignated Fleet Composite Squadron Eight (VC-8) on 18 June 1965.

At right, three VU-8 FJ-3D/DF-1D drone control aircraft. (via Mark Aldrich)

UTILITY SQUADRON TEN, VU-10 "PROUD PELICANS"

Fleet Utility Squadron Ten was originally established as VJ-16 on 1 December 1943. Stationed at NAS Guantanamo Bay, Cuba, the squadron was redesignated VU-10 on 15 November 1946. The unit had the dual mission of providing normal Atlantic Fleet utility services and providing base defense for Gitmo.

The squadron's first jets were the F9F Cougar, which were in use from 1954 through 1958 when the FJ-3s were acquired. The Furys provided the Atlantic Fleet with real-world threat training for their CIC/radar operators, air controllers and anti-aircraft weapons systems.

For base defense, the FJ-3 was assigned to the unit and four aircraft were kept on alert status. Two were maintained at alert condition five (airborne in 60 minutes) and two were held in reserve.

The FJ-3s were replaced with F-8 Crusaders before the squadron was redesignated VC-10 on 1 July 1965.

Below, VU-10 FJ-3 BuNo 136059 JH/46. (USN) Bottom, VU-10 FJ-3 BuNo 136080 at Leeward Point, Guantanamo Bay, Cuba. (Ginter collection)

Above, six VU-10 FJ-3s at Guantanamo Bay with thirteen VMF-312 FJ-3s in 1959. (USN) Below, VU-10 FJ-3 BuNo 131442 over McCalla Field, Guantanamo Bay, on 6 February 1962. Aircraft has da-glo red nose and tail and wing panels that extend from the wing-walk to the tip. (USN)

FIGHTER SQUADRON TWENTY - ONE, VF-21 "MACH BUSTERS"

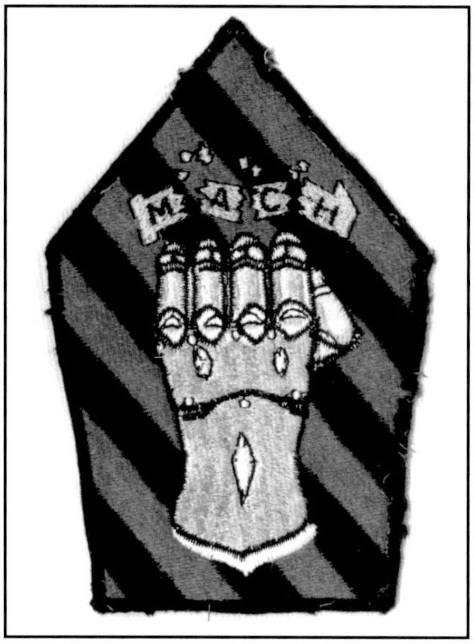

VF-21 was established as VF-74A on 1 May 1945 and redesignated VF-74 on 1 August 1945. VF-74 was redesignated again as VF-1B on 15 November 1946 before becoming VF-21 on 1 September 1948. During the Korean War the squadron flew the F9F-2/5 Panthers and transitioned to the F9F-7 Cougar in July 1953. In 1954, the F9F-6s replaced the F9F-7s and while assigned to ATG-181 deployed aboard the USS Randolph (CVA-15) from November 1954 through June 1955.

VF-21 then transitioned to the FJ-3 and became the most photographed FJ-3 squadron ever when it deployed as the lead squadron during the shakedown cruise of America's first supercarrier the USS Forestal (CVA-59) from 24 January through 31 March 1956.

ATG-181 and VF-21 deployed its FJ-3s aboard the USS Bennington (CVA-20) from 3 October 1956 through 23 May 1957. During the cruise, one Fury I/211 was lost with its pilot when it missed all the wires, went inverted before crashing on deck and falling into the sea.

The squadron transitioned to the Grumman F11F-1 Tiger in June 1957 and was redesignated VA-43 on 1 July 1959.

Above, blue VF-21 FJ-3 in flight with a VF-41 F2H-3 on 29 August 1955. (National Archives) Below, yellow and white-trimmed VF-21 FJ-3 BuNo 135862 on 6 October 1955 at NAS Oceana. (via Bob Lawson)

VF-21 "MACH BUSTERS"

Above, VF-21 CO CDR W.M. Harnish made the second landing on the USS Forrestal (CVA-59) at 2:45 pm on 3 January 1956. (USN) At right, Air Task Group 181 (ATG-181) CAG , CDR Ralph L. Werner, being congratulated by the USS Forrestal's Commanding Officer CAPT Roy Lee Johnson after making the first landing on CVA-59 at 2:40 pm on 3 January 1956. The aircraft were trimmed in yellow/orange outlined in black. The interior of the ammo door and the ammo cans were silver. The port side ammo door when opened serves as the first step for cockpit entry. (USN) Below, VF-21 FJ-3 I/207 searches for a wire on the Forrestal in February 1956 during the carrier's shakedown cruise. (National Archives)

At right, the skipper of VF-21, CDR W.M. Harnish, and the CAG of ATG-181, CDR Ralph L. Werner, discuss the Fury on deck of the Forrestal CVA-59. (USN) Below, the pilot of VF-21's FJ-3 BuNo 136128 rides his aircraft down to the hangar deck on the Forrestal in 1956. (USN)

FIGHTER SQUADRON TWENTY - ONE, VF-21 "MACH BUSTERS"

Above and below, VF-21 hangar deck spaces and stowage aboard the USS Forrestal (CVA-59) in 1956. (USN)

Above, three VF-21 FJ-3s during a catapult launch cycle. (USN) Below, VF-41 F2H-3 Banshee launches while a VA-86 F7U-3 Cutlass and a VF-21 FJ-3 Fury wait their turn. (National Archives)

LEEWARD POINT GUANTANAMO BAY, CUBA

Above, VF-21 boresights its cannons at Leeward Point Field, Guantanamo Bay, Cuba, during the shakedown cruise of the USS Forrestal. (NAA) At left, VF-21 pilots being briefed by their skipper, CDR W.M. Harnish, in their assigned ready room at Leeward Point. (NAA) Below, refueling and maintenance being performed at Leeward Point between gunnery training flights in 1956. (NAA)

Above, four VF-21 FJ-3s from the USS Bennington (CVA-20) fly by Mt. Fuji in 1957. (National Archives) Below, VF-21 FJ-3 BuNo 136123 on display at NAS Glenview, IL, on 8 July 1956. (USN)

VF-21 plane captains prepare their Furies for a mission from the USS Bennington (CVA-20). (NAA)

FIGHTER SQUADRON TWENTY- FOUR, VF-24 "CORSAIRS"

The squadron was originally established as VB-74 on 1 May 1945, with Curtiss SBW-4E Helldivers. It was redesignated VA-1B on 15 November 1946 and equipped with AD-1 Skyraiders in July 1947. It became VA-24 on 1 September 1948. F4U-4 Corsairs were received in February 1949 and the squadron's designation was changed once again on 1 December 1949, this time to VF-24. The squadron was transferred to NAS Alameda, CA, from NAAS Oceana, VA, in June 1950 in preparation for the Korean conflict.

VF-24 returned from a Korean combat cruise aboard the USS Boxer (CV-21) on 9 June 1951 and moved to NALF Santa Rosa, CA, where they transitioned to the Grumman F9F-2 Panther. After returning from their third combat tour, VF-24 became the first West Coast squadron to acquire the F9F-6 Cougar.

The FJ-3/3M Fury replaced the Cougars in late 1955. The squadron conducted one cruise with their nimble Furys. It was aboard the USS Shangri-La (CVA-38) from 13 November 1956 through 20 May 1957.

VF-24 received their first F3H-2M Demons in the summer of 1957 and retained five FJ-3s through 30 September 1957. The F11F-1 Tiger was acquired on 23 December 1958 and on 9 March 1959, VF-24 was redesignated VF-211. The squadron

Above, one of VF-24's first FJ-3s was blue with red trim and a white fuselage stripe. (via John Elliott) Below, red-trimmed VF-24 FJ-3M BuNo 136011 at Miramar, CA, on 1 September 1956. (H. G. Martin)

acquired the F8U-1 Crusader in December 1959 and flew various versions until December 1975 when the F-14A Tomcat took over.

Above, red-trimmed VF-24 CAG bird FJ-3M BuNo 136155 at NAS Miramar, CA, on 1 September 1956. Rudder stripes were: green, yellow and blue. (Peter M. Bowers) At left, VF-24's Commanding Officer prepares to launch from the Shangra-La on 29 August 1956 in BuNo 139212. (National Archives) Below, CO's bird being towed past the island on CVA-38 in January 1957. (National Archives)

FIGHTER SQUADRON THIRTY- THREE, VF-33 "TARSIERS"

Above, white-trimmed VF-33 FJ-3 piloted by Fred Martin over Cherry Point in 1955. (via RADM Tussot/Tailhook) Bottom, early slatted-wing VF-33 FJ-3 BuNo 135831 flown by LCDR Manby at NAS Norfolk, VA. (NAH)

VF-33 was established on 12 October 1948 at NAS Quonset Point, R.I. The squadron was first equipped with Grumman F8F-2 Bearcats. The unit transitioned to the Chance Vought F4U-4 Corsair and deployed once to the Mediterranean aboard the USS Kearsarge (CVA-33) and once to Korea aboard the USS Leyte (CVA-32).

In May 1953, the squadron transitioned to the Grumman F9F-6 Cougar. While assigned to CVG-6, the unit deployed its Cougars to the Mediterranean aboard the USS Midway (CVA-41) under the command of CDR K.H. Stefan. VF-33 participated in NATO Operation Hellenic Sky 1 in February 1954 and Italic Sky 1 in April 1954.

After the deployment, the squadron moved to NAS Oceana in September 1954 and received the North American FJ-3 Fury in November. On 27 January 1955, LCDR W.J. Manby set a new unofficial climb record by reaching 10,000 feet from a standing start in 73.2 seconds at NAS Oceana.

In the spring of 1955, the squadron made several trips to Guantanamo Bay for training and gunnery practice and aboard the USS Ticonderoga for carrier qualifications. VF-33 deployed to the Mediteranean aboard the USS Lake Champlain (CVA-39) from 9 September 1955 through 31 March 1956 while under the command of CDR R.H. Jester.

After returning to Oceana, VF-33 won the Atlantic Fleet Gunnery Championship. Due to this, during the 1956 Fleet Air Gunnery Meet, at El Centro, AIRLANT sent VF-33 to represent it in the day fighter category which was won by VF-212, an AIRPAC F9F-8 squadron. In April 1957, VF-33's FJ-3s were again sent to the Naval Air Gunnery Meet to represent AIRLANT.

After returning, the squadron began receiving FJ-3Ms and fielded a mixed bag of FJ-3s and FJ-3Ms for a Carribean deployment abord the USS Intrepid (CVA-11). The cruise was from 20 May through 29 July 1957. Sidewinder air-to-air missile firing

Above, Greek officers from the Greek War College pose in front of VF-33 CO's Fury on CVA-39 on 7 March 1956. (National Archives) At left, FJ-3 BuNo 135899 at New Castle, DE, in May 1956. (Ron Picciani) Bottom, VF-33 FJ-3 C/203 at NAS Miami, FL, in May 1955. Note MIAMI on the tail of the Marine F3D Skyknight at right. (Clay Jansson)

training exercises were carried out from 1 July through 20 July 1957 off

Guantanamo Bay.

After returning to Virgina, the squadron remained at Oceana for only a week before conducting a short at-sea deployment on CVA-11 from 10 August through 15 August and again from 27 August through 22 October 1957. VF-33 took part in Operation Sea Spray from 9 September through 12 September where they conducted Combat Air Patrol with NATO forces while enroute to the United Kingdom. This was followed by Operation Strike Back from 19 September through 28 September 1957 where they provided Combat Air Patrol for NATO forces while operating in the Norwegian Sea. CAP was performed again from 12 October through 15 October for NATO Operation Pipedown and again from 16 October through 21 October during Operation Arnex.

From 4 November through 13 November 1957, VF-33 participated in aerobatic manuevers, strafing and Sidewinder missile firing during a Armed Forces Staff College Demonstration Cruise.

Once back at Oceana, the squadron conducted instrument training through the 1st of December 1957 when transition training began on the new Grumman F11F-1 Tiger. The squadron had ten FJ-3/3Ms and thirteen Tigers on hand on 31 December.

The F11F-1 Tiger was replaced by the Vought F8U-1 Crusader in February 1961. In November 1964, the unit transitioned to the McDonnell F-4 Phantom which it traded in for F-14 Tomcats in July 1981. The squadron was disestablished on 24 September 1993.

At top, VF-33 FJ-3 BuNo 135844 at St. Louis. (Bill Rudolph via Fred Roos) Above and at right, Kodak Brownie photos of VF-33 FJ-3s on the USS Lake Champlain (CVA-39) as it pulls into a Mediterranean port in 1956. (Ginter collection)

Above, VF-33 FJ-3 BuNo 141418 belonged to CO CDR J.V. Ferris. It undergoes maintenance during the second Naval Weapons Meet, El Centro, CA, in April 1957. (National Archives) Below, three VF-33 FJ-3s BuNos 141418 C/201, 139247 C/214 and 141426 C/206 in flight over El Centro on 5 April 1957. (National Archives)

Above, VF-33 FJ-3Ms while assigned to CVG-6 in 1957 aboard the USS Intrepid (CVA-11). Trim was orange/yellow bordered by black. Intake lips were red. BuNo 141422 C/203 is in the foreground. (USN) Below, VF-33 FJ-3 BuNo 141418 belonged to CO CDR J.V. Ferris at the second Naval Air Weapons Meet in April 1957. Work is being conducted on the aircraft's fire control system (radar computer) by AQ-1 Gossman. Oxygen cylinder is visible in the lower open access door. (National Archives)

FIGHTER SQUADRON FIFTY - ONE, VF-51 "SCREAMING EAGLES"

The Screaming Eagles started out in 1927 as VF-3S flying the Curtiss F6C-4. The squadron designation was soon changed to VF-3B and the Boeing FB-5 replaced the F6C-4. In 1929, the FB-5 gave way to the Boeing F3B-1 which was traded for the Boeing F4B-4 in 1931. In 1935 the retractable gear Grumman F2F-1 joined the squadron. In 1937 the squadron was redesignated VF-5B and upgraded to the Grumman F3F-3. When war broke out on 7 December 1941, the squadron was equipped with Grumman F4F-3 Wildcats.

VF-5B was disestablished on 7 January 1943, and most of the squadron's assets including personnel were assigned to the newly-established VF-1. Flying Grumman F6F Hellcats, VF-1 was redesignated VF-5 on 15 July 1943. The squadron finished out the war in Vought F4U Corsairs and was redesignated VF-5A on 15 November 1946 while flying the Grumman F8F-1 Bearcat. The Bearcat was replaced by the North American FJ-1 Fury on 18 November 1947 and VF-5A became the first Navy fleet squadron to operationally deploy jets at sea.

The Screaming Eagles were once again redesignated on 16 August 1948, when they were given their current designation of VF-51. Grumman F9F-3 Panthers replaced the Furies and the squadron flew F9F-2/2Bs, F9F-3s and F9F-5s during three Korean War deployments.

Above, pre-delivery North American publicity illustration of a VF-51 FJ-3. (Ginter collection) Below, VF-51 FJ-3 BuNo 135829 piloted by LCDR Alex Varicu in 1955 at NAS North Island, CA. (Alex Vericu via Tailhook)

In late 1953, F9F-6 Cougars replaced the Panthers, which were replaced with North American FJ-3 Furies in December 1954. The squadron deployed its Cougars aboard the USS Philippine Sea (CVA-47) during 1954.

VF-51 was the first Navy fighter squadron to receive the FJ-3 and was responsible for developing much of its tactical usage.

During the April 1957 Naval Air Weapons Meet at NAAS El Centro, CA, VF-51's CO, WWII Ace CDR A. Vraciu racked-up the best individual score of the meet. The squadron completed the competition in second place.

Above, VF-51 FJ-3 BuNo 136998 at NAS Miramar, CA, on 30 October 1955. (William Swisher) Below, four VF-51 FJ-3s in flight in 1955. The three fuselage stripes were white and the intake lip and horizontal fuselage stripe was red. (USN via Thomas Gates)

VF-51 deployed with fourteen of their FJ-3/3Ms aboard the USS Bon

Above, eight VF-51 FJ-3s in flight off San Diego with VF-142 FJ-3s in 1956. (USN) Below, tail section of VF-51 FJ-3 BuNo 135965 being removed for an engine inspection in 1956. (USN via Clay Jannson)

Above, VF-51 Fury BuNo 136017 at NAS Miramar, CA, in 1956. (USN via Thomas Gates/Tailhook) At right, VF-51 1957 Fleet Air Gunnery Meet pilots. (via Mark Aldrich) Below, four VF-51 FJ-3s, BuNos 136156 NF/101, 139245 NF/104, 139265 NF/110 and 136153 NF/112 in flight over NAAS El Centro during the second annual Fleet Air Gunnery Meet on 2 April 1957. (USN)

Homme Richard (CVA-31) from 10 July 1957 through 11 December 1957. On 30 July, an Operational

Readiness Inspection (ORI) was performed and the squadron received the highest score in Air Group Five with an overall grade of 92.61.

Above, VF-51 FJ-3 BuNo 136162 during the Fleet Air Gunnery Meet at El Centro in April 1957. (via Tailhook) Below, BuNo 139245 returning to El Centro. (USN) Bottom, LCDR Blackman in BuNo 136153. (USN)

Above, VMA-211 Skyraider being armed with three VF-51 FJ-3s in the background on 2 April 1957. (USN) Below, VF-51 FJ-3 landing on the USS Bon Homme Richard (CVA-31). CO CDR A. Vraciu was inset into the photo. (USN)

After the cruise only nine FJ-3s remained on hand at NAS Miramar as of 31 December 1957. Aircraft on hand on 30 June 1958 were three FJ-3s and nine F11F-1s.

The squadron completed its transition to the supersonic Grumman F11F-1 Tiger in 1958 only to re-equip with Douglas F4D-1 Skyrays in January 1959. On 10 November 1960, the squadron transitioned to the Vought F8U Crusader. The squadron spent ten years in Crusaders and transitioned to the McDonnell F-4B Phantom II in November 1970. F-4Ns were acquired in early 1974 and the F-14A Tomcat was acquired on 16 June 1978. The squadron was disestablished on 16 February 1995.

At left, VF-51 pilot mans his plane while maintenance is conducted on VF-141 F4D-1 Skyrays in 1957. (via Mark Aldrich) Below, VF-51 FJ-3 BuNo 135908 on CVA-31 in 1957. (via Mark Aldrich) Above right, VF-51 FJ-3 Fury BuNo 136051 taxis past its squadron mates on the deck of CVA-31 in 1957. (via Mark Aldrich) At right, VF-51 Furies spotted on CVA-31's foredeck.

Above, VF-51 Furies share the deck with a VF-141 F4D-1 Skyray on CVA-31 in 1957. (USN) At right, the Grumman F11F-1 Tiger began arriving in November 1957. An unmarked Tiger is seen here ready to launch on a familiarization flight with a squadron FJ-3M. (USN) Below, VF-51 FJ-3M BuNo 135844 belonging to the squadron's Commanding Officer CDR A. Vraciu is positioned on the deck of CVA-31 in October 1957. (USN)

FIGHTER SQUADRON FIFTY - THREE, VF-53 "BLUE KNIGHTS"

In 1948, under the command of LCDR W. D. Hubbell, the squadron was organized as part of Air Group Five with Grumman F8F Bearcats. 1949 brought a new Commanding Officer, LCDR W. R. Pittman, and a new aircraft, the Chance Vought F4U Corsair. The squadron was in the Far East aboard the USS Valley Forge (CV-45) when the Korean War broke out. The ship was ordered to Korea in June1950 and launched its first strike on 3 July. The squadron returned to San Diego in November 1950.

A second Corsair war cruise took place aboard the USS Essex (CV-9) from June 1951 through April 1952. CDR Trum was relieved by LCDR Lawrence B. Green at NAS San Diego and the squadron transitioned to the Grumman F9F-2 Panther.

In the summer of 1952, F9F-5s replaced the F9F-2s. Two Korean War deployments were made while flying Panthers. In 1955, VF-53 replaced its Panthers with F9F-8 Cougars and in 1957, the FJ-3M Fury took over.

Above, VF-53 CAG in flight. (USN) Below, VF-53 FJ-3Ms BuNos 136153 NC/112 and 139225 NC/113 off the USS Kearsarge (CVA-33) over the Pacific. (USN)

VF-53 deployed their FJ-3Ms aboard USS Kearsarge (CVA-33) as part of ATG-3 from 9 August 1957 through 2 April 1958. Prior to their deployment to WestPac the squadron received the highest score during the

Above, green clad VF-53 pilots carry or wear their yellow Mae Wests prior to a flight from CVA-33 in 1957. Left-to-right; Burr, Person, Fields, Andrews, Blackman, Nichol, Pareau, Terry, unknown, Smyer and Albright. (Dick Murdock via Tailhook) Below, VF-53 FJ-3M BuNo 136150. (via Norm Taylor)

Above, VF-53 FJ-3Ms share the deck with a VFP-61 photo Cougar. BuNo 139225 is in the foreground. (via Norm Taylor) At right, two VF-53 Furies over CVA-33. (USN) Below, red-trimmed BuNo 135990 on the deck of CVA-33 in 1957. (USN)

Air Wing's Operational Readiness Inspection.

The squadron was redesignated VF-124 on 11 April 1958 and equipped with F8U-1 Crusaders. VF-124 transitioned to F-14A Tomcats and was disestablished on 30 September 1994.

FIGHTER SQUADRON SIXTY - ONE, VF-61 "JOLLY ROGERS"

The second VF-17 was established on 1 August 1944 with the Grumman F6F Hellcat. The squadron transitioned to Chance Vought F4U-4 Corsairs in late 1945 and was redesignated VF-5B on 15 November 1946. VF-5B was redesignated VF-61 on 28 July 1948 and transitioned to Grumman F8F-2 Bearcat. The Grumman F9F-2 Panther was acquired in April 1950.

In late 1953 the squadron transitioned to the F9F-6 Cougar. The FJ-3 Fury was received in mid-1955, only to be replaced with the F9F-8 in late 1955. Carrier qualifications with the FJ-3s were conducted aboard the USS Lake Champlain (CVA-39) in June 1955.

The Cougars were replaced with F3H-2M Demons in September 1956. In 1959, the squadron was scheduled to receive the F8U-2 Crusader, but instead was disestablished on 15 April 1959. In June 1959, VF-84 took over the Jolly Roger name, insignia and traditions.

At top, white-trimmed VF-61 FJ-3 E/206 in June 1955. (Bob Lawson via Tailhook) At right, VF-61 FJ-3 E/215 in 1955. Dark blue aircraft was trimmed in white with distinct Jolly Roger insignia on the fuselage side. (USN)

Next page top, VF-61 flightline in 1955. (via Mark Aldrich) Bottom, VF-62 FJ-3 BuNo 136140 in 1955. (via M. Aldrich)

215
NAVY
E

206
NAVY
X
06

FIGHTER SQUADRON SIXTY - TWO, VF-62 "BOOMERANGS"

VF-62 was established on 1 July 1955 at NAS Cecil Field, FL, where they equipped with the FJ-3/3M Fury. The Boomerangs took part in two deployments aboard the USS Randolph (CVA-15) as part of Air Task Group 202 (ATG-202). The first was the shakedown cruise of Randolph to the Carribiean from 1 January to 9 March 1956. The second deployment was to the Mediterranean

Above, VF-62 FJ-3 BuNo 136138 landing aboard the USS Randolph (CVA-15) in November 1956. Aircraft had white stars on red tail stripes and intakes. (USN) Below, three VF-62 FJ-3s share the Randolph's deck with an AJ-2 in 1956. (Norm Taylor archives) At top right, VF-62 FJ-3 BuNo 135897 landing on CVA-15. (via Mark Aldrich) At right, VF-62 red-trimmed FJ-3 being serviced after a gunnery flight in 1955. (N.Taylor archives) Below right, two VF-62 FJ-3M Furies BuNos 139232 AP/209 and 139231 AP/207 from the USS Essex (CVA-9) refueling from a VAH-7 AJ-2 on 15 June 1958. (USN)

X
5897
201
NAVY
NC
109 X
110
NAVY

X
206
NAVY
06

AP
207
NAVY
AP
134043
NAVY
VAH-7
501
AP
9232
209
NAVY
209AP

from 14 July 1956 through 19 February 1957 where VF-62 took part in the Suez Crisis.

CDR Roman Kolakowski was in command during the squadron's third deployment aboard the USS Essex (CVA-9) from 2 February through 17 November 1958 for a Med-Far East

At right, VF-62 FJ-3M on Essex being loaded with Sidewinders in July 1958 during the Lebanon Crisis. (USN) Below, four VF-62 FJ-3Ms BuNos 139238 AP/211, 139227 AP/205, 139231 AP/207, and 139230 AP/208 in flight from the USS Essex (CVA-9) in 1958. The top aircraft has a yellow fuselage stripe just aft of the cockpit and a boomerang-shaped yellow marking on the tail. (National Archives)

cruise as part of ATG-201. While in the Med the squadron took part in the Lebanon Crisis and conducted operations to protect Quemoy in the Western Pacific.

Two more Essex deployments were conducted as part

Above, VF-62 CAG bird with ATG-202 "AQ" tail code. (via Mark Aldrich) Below, VF-62 FJ-3M BuNo 139232 refuels from AD Skyraider while armed with Sidewinders during the Lebanon crisis. The landing gear on the Fury had to be extended to slow it down enough to tank from the slower AD. (via Mark Aldrich)

of CVG-10. A short cruise into the Atlantic was made from 2 May through 21 July 1959, and a full-length Med cruise was conducted from 7 August 1959 through 26 February 1960.

After the last Essex deployment

Above, VF-62 CAG bird FJ-3M BuNo 139227 in 1958 with yellow and black trimmed tail and drop tank fins. (Ginter collection) At right, VF-62 FJ-3M BuNo 139211 with "AK" tail code of CVG-10 traps aboard the USS Essex in 1960. (via Mark Aldrich) Below, VF-62 FJ-3M BuNo 139231 in flight over Beirut in July 1958. (Norm Taylor archives)

VF-62 transitioned to the F8U-1 Crusader starting on 1 March 1960. They continued to fly Crusaders until they were disestablished on 1 October 1969.

At top, Essex-based VMA-225 A4D-2 refuels VF-62's CAG bird FJ-3M BuNo 136137 on 25 January 1960. Note squadron insignia on the forward fuselage and the Carrier Air Group insignia on the tail. (USN) At right, VF-62 pilots in 1959. Front row L/R; LT Pollard, LT Almberg, CDR Stone, CDR Winslow, LCDR Shea, LT Lambert. 2nd row: Jones NAA, LTJG Patterson, LTJG Berg, LTJG Query, LTJG Porter, LTJG Donovon, LT Rose. 3rd row: LTJG Kaufman, LTJG Snow, LTJG Fowlkes, LTJG Brown, LTJG Miller, LTJG Chalfant. Bottom, retired VF-62 Fury BuNo 135822 at NAF Litchfield Park on 21 March 1960. (William Swisher)

FIGHTER SQUADRON SEVENTY - THREE, VF-73 "BOOMERANGS"

VF-73 was established on 27 July 1948 at NAS Quonset Point, RI, with F8F-2 Bearcats. In the spring of 1949, the F8F-2s were replaced by F8F-1s. F4U-4 Corsairs started replacing the Bearcats in the spring of 1950 and the squadron flew a mixed bag of Corsairs and Bearcats until F9F-5 Panthers were received in November of 1951. In December 1952, F6F-5 Hellcats were temporarily acquired to replace the F9F-5s until another front-line aircraft could be found to equip the squadron. In 1953, VF-73 acquired the F2H-3 all-weather Banshee for a short period prior to acquiring the F9F-6 Cougar.

Above, VF-73 FJ-3 leaves the deck of the USS Randolph CVA-15 on 22 March 1957. (National Archives) Below, VF-73 FJ-3 at NAS Quonset Point in October 1956 with VU-2 target tugs in the background. (National Archives)

Above, VF-73 FJ-3 BuNo 135942 on 8 July 1956. Trim was light blue equivalent to powder blue. (via Mark Aldrich) Below, VF-73 FJ-3M BuNo 141435 after trapping aboard the USS Randolph (CVA-15) on 22 March 1957. (National Archives)

At right, VF-73 FJ-3 BuNo 136073 at an Air Show in 1957. (via Tailhook) Below right, VF-73 FJ-3M BuNo 141427 on the Randolph. (via Mark Aldrich) Bottom, flight of four VF-73 FJ-3Ms on 8 July 1956. BuNos 141440 AD/310, 141437 AD/309, 136106 AD/306, and 136143 AD/302. (USN via Mark Aldrich)

Cougar carrier qualifications were conducted aboard the USS Tarawa (CVA-40) in December 1954 and aboard Hornet in April 1955. VF-73 deployed aboard the USS Hornet (CVA-12) from 4 May through 10 December 1955 during the ship's last straight-deck cruise.

The Cougars were flown until 1956 when the FJ-3M Fury was received. Quonset Point-based VF-73 deployed its FJ-3Ms aboard the USS Randolph (CVA-15) from 8 June 1957 through 23 February 1958. The squadron conducted operations in the Mediterranean as part of CVG-4. The squadron flew the Furies until its disestablishment on 1 March 1958.

FIGHTER SQUADRON EIGHTY - FOUR, VF-84 "VAGABONDS"

VF-84 was established on 1 July 1955 at NAS Oceana, VA, and was equipped with the FJ-3/3M Fury.

The squadron made two short deployments as part of ATG-182 aboard the USS Lake Champlain (CVA-39). The first was to the Caribbean for operational and weapons training from 10 October to 14 November 1956. The second was a sortie into the Atlantic from 15 November through 10 December 1956 in response to the Suez Crisis.

VF-84 made two deployments as part of CVG-1 aboard America's new super carrier, the USS Forrestal (CVA-59) in 1957. In response to increasing tensions in the Middle East, she deployed on 15 January 1957 for duty with the Sixth Fleet in the Mediterranean. The most powerful warship/Air Group in the world projected its strength and capabilities during the Jordanian Crisis by visiting ten ports and putting on numerous air power displays. Ports-of-call were Gibralter, Barcelona, Cannes,

At top, VF-84 FJ-3 BuNo 139250 taxis on Forrestal in May 1957. (National Archives) Middle, two VF-84 FJ-3Ms, BuNos 139235 O/206 and 139232 O/208, in launch position on the USS Forrestal (CVA-59) on 24 July 1957. (National Archives) At right, VF-84 FJ-3 BuNo 139224 ready for launch with a VA-76 F9F-8 Cougar in the background in May 1957. (National Archives)

Above and below, VF-84 FJ-3Ms BuNos 139230 O/210, 139234 O/207, 139231 O/209 and 139232 O/208 in flight in 1957. Fuselage and tail trim was yellow and black. (USN/National Archives)

Taranto, Naples, Rome, Athens, Rhodes, Istanbul, and Beirut. The squadron participated in five NATO exercises, three national operations and a British-American exercise. The deployment ended on 22 July 1957.

After a short yard period the Forrestal sailed again to take part in NATO exercise Operation Strikeback in the North Atlantic from 16 August to 14 November 1957. This was the largest NATO exercise ever held until that date.

The squadron's FJ-3Ms made a final deployment during the Lebanon Crisis as part of CVG-7 aboard the USS Randolph (CVA-15) from 3 July 1958 through 12 March 1959.

In April 1959, the Vagabonds began receiving the F8U-2 Crusader. The last Fury was transferred out on 24 June 1959. After receiving the Crusaders, the squadron took over the heritage and insignia of the Jolly Rogers after the disestablishment of VF-61. The Phantom arrived in 1964 and the Tomcat in 1976. VF-84 was disestablished on 1 October 1995.

Above, four VF-84 FJ-3Ms launching from the USS Forrestal (CVA-59). (USN) Below, two VF-84 FJ-3Ms, BuNos 139250 O/202 and 139227 O/211, taxi into launch position on CVA-59 on 24 July 1957. (National Archives)

Above, VF-84 FJ-3Ms BuNos 139238 O/204, 139227 and 139232 O/208 share the deck with a VF-14 F3H-2M in late 1956. (USN) Below, missile-armed VF-84 FJ-3Ms being prepped for launch from CVA-59 in May 1957. (National Archives)

VF-84 "VAGABONDS"

Above, VF-84 FJ-3M BuNo 139234 on CVA-59 with new AB tail code in October 1957. (via Burger) At right, VF-84 FJ-3M BuNo 139237 aboard the USS Forrestal (CVA-59) during Operation Strike Back in September 1957. (National Archives) Below, VF-84 Vagabond FJ-3M BuNo 139234 being pushed back after a trap on CVA-59. (USN)

Above, yellow-trimmed VF-84 FJ-3M BuNo 141422 with new CVG-7 "AG" tail code. (via Mark Aldrich) Below, VF-84 FJ-3M launching from the USS Randolph. (USN) Bottom, the new and the old! VF-84 FJ-3M BuNo 141415 flanked by the new F8U-2.

FIGHTER SQUADRON NINETY - ONE, VF-91 "RED LIGHTNINGS"

VF-20 was established in 1942 with F6F-3 Hellcats. In August 1944, F6F-5s replaced the F6F-3 Hellcats. Redesignated VF-9A on 15 November 1946, the squadron flew F8F-1 Bearcats and F6F-5P Hellcats from CV-47. In September 1948, VF-9A was redesignated VF-91 and the Grumman F9F-2 Panther was acquired in 1952.

VF-91 took their Panthers aboard the USS Philippine Sea (CVA-47) for a Korean War deployment from 15 December 1952 through 14 August 1953. After returning to CONUS, the squadron transitioned to the Grumman F9F-6 Cougar in 1954.

In the summer of 1956, the North American FJ-3 Fury replaced the Cougar. During VF-91's weapons deployment to Fallon, NV, in the spring of 1957, the squadron recorded the highest percentage of fire-outs ever obtained by an FJ-3 unit. Loading 40,942 rounds of 20mm during the deployment, they fired-out 34,399 rounds or 83.8%. During a two-day competition a 92.9% fire-out was logged. The squadron was based at NAS Alameda, CA, and was commanded by CDR T.A. Turner.

The squadron took its FJ-3Ms aboard the USS Ticonderoga (CVA-14) to the Western Pacific as part of CVG-9 from 16 September 1957 to 25 April 1958. Ports-of-call were Hawaii, Japan, Subic Bay, Hong

Above, VF-91 FJ-3 BuNo 135901 on the USS Ticonderoga (CVA-14) in 1957. (USN) Below, red-trimmed VF-91 FJ-3 BuNo 135845 on the USS Ticonderoga (CVA-14) in 1957. Aircraft trim was red. (USN)

Above, VF-91 FJ-3 BuNo 135845 on the USS Ticonderoga (CVA-14) in 1957. Aircraft trim was red. (USN) At left, VF-91 FJ-3 BuNo 135989 on the USS Ticonderoga (CVA-14) in 1957. (USN) Below, two VF-91 FJ-3s, BuNos 135943 and 135909, in 1958. Note red wing tip markings. (Ginter collection)

Kong and Okinawa.

The squadron went supersonic in December 1958 with the arrival of Vought F8U-2 Crusaders. On 1 August 1963, VF-91 was redesignated VF-194. As VF-194, the squadron flew the F-8C, F-8E, F-8B, and F-8J. The McDonnell F-4J Phantom II started arriving in March 1976 and was flown until the squadron was disestablished on 1 March 1978.

FIGHTER SQUADRON NINETY - FOUR, VF-94 "TOUGH KITTY"

Fighter Squadron Ninety-Four (VF-94) was established on 26 March 1952 with F4U Corsairs which it flew during the Korean War. The squadron was based at NAS Moffett Field, CA, when it received the new FJ-3 Fury. Due to serious problems with the new

Below, two blue VF-94 FJ-3s with original "N" tail code in flight in August 1955. Lightning bolts were red. (Ginter collection) Bottom, three VF-94 FJ-3Ms on the deck of the USS Hornet in 1958. (USN)

Above and below, VF-94 operations aboard the USS Hornet (CVA-12) in 1958. (USN)

J65 engine, the FJ-3s were quickly replaced with F9F-8/8B Cougars in time for the squadron to make a WESTPAC deployment in early 1956.

VF-94 deployed to the Western Pacific aboard the USS Yorktown (CVA-10) as part of ATG-4 from 19 March through 13 September 1956. The Cougars remained until November 1957 when the FJ-3/3M was once again assigned to the unit.

VF-94 while commanded by CDR F.X. Brady made one deployment with the FJ-3/3M. It was to the Western Pacific aboard the USS Hornet (CVA-12) from 6 January 1958 through 2 July 1958. Two squadron pilots were lost during the cruise: LTJGs Alvan D. Sullivan and Robert R. Briggs. After the 1958 deployment, VF-94 was redesignated Attack Squadron Ninety-Four (VA-94) on 1 August 1958. As VA-94, the squadron transitioned to the FJ-4

Above, VF-94 FJ-3M rides the elevator on the USS Hornet (CVA-12). (USN) Below, the Hornet's deck populated by ADs, FJ-4s, F2H-3s and a VF-94 FJ-3M Fury. (USN)

Fury. In January 1959, A4D-2s replaced the Furies. Then, in September 1960, A4D-2Ns were acquired to replace the A4D-2s. On 23 October 1967, the A-4E replaced the squadron's A-4Cs and on 16 November they received their first "Super Echo". The squadron transitioned to the Vought A-7E Corsair in January 1971. The Corsairs were replaced with F/A-18C Hornets in May 1990 and the squadron was redesignated VFA-94 on 28 June 1990.

FIGHTER SQUADRON ONE - ELEVEN, VF-111 "SUNDOWNERS"

Above, Dick Mudgett and fellow pilot pose with a squadron Fury with ATG-1 fuselage markings in 1958. (Dick Mudgett via Mark Aldridge) Below, Fleet Air Gunnery Meet winners CDR F.E. Miller, LT P.H. Speer, LTJG H.L. Landry, LTJG R. Mudgett and LTJG D. Macintyre. (Dick Mudgett via Tailhook)

VF-111 was established as VF-11 on 10 October 1942. The squadron was redesignated VF-11A on 15 November 1946 and finally VF-111 on 15 July 1948.

During World War Two, the unit flew the F4F Wildcat and the F6F Hellcat. After the war, VF-111 returned to NAS North Island and transitioned to the F8F-1 Bearcat. In 1950, the F9F-2 Panther was acquired and fought well during the squadron's second war, Korea.

The F9F-6 Cougar was acquired in 1954 and the F9F-8 in 1956. After a three-week deployment to NAAS Fallon, NV, ending on 19 July 1957, the squadron transitioned to the FJ-3/3M Fury. The transition took place at NAS Alameda, CA, during which VF-111 participated in Operation Squeeze Play, a National Air Defense exercise. This was followed by Operation Fireball, a two-week air-to-air gunnery and strafing deployment to NAAS Fallon, NV. Carrier Qualifications were conducted off San Diego aboard the USS Bennington (CVA-20) from 29 November through 12 December 1957. Another deployment to Fallon for air-to-air gunnery, called Operation Snow Ball, occurred from

Four Sidewinder-equipped VF-111 FJ-3Ms, BuNos 141434 NA/113, 141399 NA/105, 141441 NA/111, and 141413 NA/112, over Oahu on 10 September 1958. When first deployed the squadron retained all ATG-1 markings including it's "NA" tail code. Further into the cruise the "NA" tail code was replaced with ATG-4's "ND" tail code but retained the "NA" code on the wings. By the end of the cruise some refinished aircraft had the ATG-1 fuselage markings removed. (USN)

13 January through 24 January 1958. This was followed by another round of carrier qualifications from 10 February through 21 February 1958.

The squadron was picked to participate in the Fleet Air Gunnery Meet and commenced practice at NAAS El Centro, CA, on 16 March 1958. Actual competition started on 14 April resulting in VF-111 winning the air-to-air day fighter category. LTJG R.L. Mudgett placed third in the individual scoring.

After winning the Weapons Meet, CDR F.E. Miller was relieved by LCDR Richard W. Huxford on 24 April. An Administrative/Material Inspection was held that same day and VF-111 received a grade of excellent. Sidewinder training then commenced and all pilots became qualified by 17 June even though twelve of fourteen FJ-3Ms were grounded due to faulty main landing gear. Two at-sea training periods were also conducted during this period; one from 26 April through 11 May and one from 22 May through 23 June.

On 5 July through 21 July 1958, VF-111 boarded the Bennington for participation in the British Columbia Centennial Celebration during which 17,000 people viewed squadron aircraft. This was followed by another Fallon training exercise from 22 July through 10 August and a short final training cycle aboard CVA-20 from 10 August through 16 August.

The actual deployment began on 22 August 1958 and concluded on 12 January 1959. An ORI, given on 9 September, found the squadron receiving an overall grade of excellent (91%). During the cruise, VF-111 participated in the Manila Air Show. Eight FJ-3Ms formed two diamonds inside of a larger diamond composed of other ATG-4 aircraft.

VF-111 was disestablished on 19 January 1959.

At top, LSO watches a VF-111 Fury trap aboard CVA-20 in March 1958. (Dick Mudgett/Tailhook) At left, two VF-111 FJ-3Ms escort a photo Cougar in 1958. (USN)

Above, four VF-111 FJ-3Ms in flight in September 1958. (USN) Below, VF-111 FJ-3M being re-spotted for maintenance aboard the USS Bennington (CVA-20) in 1958. Note red trim on the fin tip and wing tips. (USN)

VF-111 FJ-3M BuNo 141441 off Oahu in 1958 with ATG-4's "ND" tail code. (USN) Below, VF-111 squadron photo aboard the USS Bennington (CVA-20) in late 1958. (via Mark Aldrich)

FIGHTER SQUADRON ONE-TWENTY-ONE, VF-121 "PACEMAKERS"

Below left, CO, CDR T.S. Sedaker, mounting his FJ-3M on the deck of the USS Lexington in 1957. (USN) Above, CO CDR Sedaker taxis BuNo 141401 at NAS Miramar, CA. (via Tailhook) Below, two VF-121 FJ-3Ms taxi past the island of CVA-16 in 1957. (Tailhook) Bottom, red-trimmed VF-121 FJ-3M BuNo 141414 taxis on the USS Lexington (CVA-16) in 1957. (Ray Sinwell)

NAS Los Alamitos reserve squadron VF-781 was established on 1 July 1946 flying F6F-5 Hellcats. The squadron volunteered for active duty 100% at the outbreak of the Korean War in July 1950. LCDR Collin Oveland moved his squadron to NAS North Island where it combined with three other fighter squadrons and one attack squadron to form CVG-102. The squadron started forming-up on the F4U-4 but transitioned instead to F9F-2 Panthers by the end of the

year. Due to the overcrowding at North Island VF-781 relocated to NAAS Miramar 20 miles away. As a result of the move, VF-781 became the first Navy fighter squadron to be assigned to Miramar after the Marines gave up the base.

On 10 May 1951, CVG-102 reported aboard the USS Bon Homme Richard (CV-31) and left North Island for the war zone.

During the second Panther cruise on 4 February 1953, while off the coast of Korea, VF-781 was redesignated VF-121.

Once back at Miramar, VF-121 under the command of CDR J. E. Savage, transitioned to the new Grumman F9F-6 Cougar. A Boxer deployment was made and then the squadron obtained F9F-8 Cougars.

The squadron began to transition to the FJ-3M Fury in March 1957 in preparation for its 1957 WestPac deployment aboard the USS Lexington (CVA-16). The cruise commenced on 19 April 1957 and ended on 18 October 1957 after visiting nine ports-of-call.

In 1958 the F11F-1 was acquired and in April the squadron became the Replacement Air Group (RAG) for both the F11F Tiger and F3H Demon. In October 1962, they became the F-4 Phantom RAG, a job they performed until disestablishment in September 1980.

Above, VF-121 FJ-3M BuNo 136151 in flight in 1957. The red lightning bolt was outlined in black. The tail stripes between the thin white stripes above and below the tail code stripe were black. (Ray Sinwell) Below, VF-121 FJ-3M BuNo 141410 in flight over Southern California in 1957. (Ray Sinwell)

Above, red trimmed VF-121 FJ-3M BuNo 141402 in flight. (CDR Art Hedberg via Tailhook) Below, four VF-121 FJ-3Ms from CVA-16 in flight on 20 June 1957. (USN) Bottom, BuNo 141402 traps with a Sidewinder mounted under the right wing in 1957. (CDR Art Hedberg via Tailhook)

FIGHTER SQUADRON ONE FORTY-TWO, VF-142 "FIGHTING FALCONS"

On 1 November 1949, VF-791 was established at NAS Memphis with F6F-5 Hellcats. VF-791 was called to active duty on 20 July 1950. By the 26th the squadron found itself at NAS San Diego, CA, where they set about transitioning to the Vought F4U-4 Corsair. A war cruise aboard the USS Boxer (CV-21) followed with the squadron returning to San Diego on 24 October 1951. The squadron received two F9F-2 Panthers in March 1952, but proceeded to be the first Pacific Fleet squadron to receive the F9F-5 Panthers.

VF-791 transferred to NAS Miramar on 4 December 1952 and on 4 February 1953 the squadron was redesignated VF-142 when reserve Air Group 101 (CVG-101) was augmented into the regular Navy as CVG-14. In May 1953, the squadron transitioned to the Grumman F9F-6 Cougar. The Cougars gave way to Furies in March 1956 when the squadron received the FJ-3M.

On 17 January 1956, while returning on Boxer from a WestPac cruise in Cougars, LCDR Ralph Hanks took ten of his crew aboard a MATS flight to Moffitt Field to attend the Jet Transitional Training Unit operated by VC-3. The pilots com-

Above, VF-142 FJ-3M BuNo 139257 with dark blue trim. (via Mark Aldrich) Below, VF-142 FJ-3M BuNo 141424 from the USS Hornet CVA-12 while assigned to CVG-14 in 1956. (USN)

Above, VF-142 FJ-3s in flight off San Diego in 1956. Although the trim colors in these photos apear to be black, color photos show them to be blue. (USN) At right, VF-142 squadron photo at NAS Miramar on 17 January 1957. Left-to-right: CDR E.R. Hanks CO, LTJG R.R. Wilmer, LCDR A.M. Adams, LTJG P.D. Hoskins, LCDR P. Bugg XO, LTJG P. Shimek, LTJG J.J.Jester, LTJG P.E. Bovey, LCDR M.C. Griffin, LTJG R.L. Merritt, LT E.R. Genier, LTJG R.G. Conaughton, LCDR B.S. Franklin, LTJG R.E. Gregory, LTJG E.L. Geronime, LTJG R.D. Nelick, LTJG A.W. Markham. (USN) Below, four VF-142 Furies over the Pacific in February 1957. (USN)

pleting training at VC-3 were LCDR Hanks, his XO LCDR Paul Bugg, LT Blaylock, LT Billy S. Franklin and LTJG Glen E. "Gabby". The squadron then ferried their new Furies direct from the factory in Columbus, OH, to NAS Miramar. In August and September the squadron qualified in gunnery and inflight refueling while operating from NAAS El Centro. During October and November the squadron was checked out on the new Sidewinder missile with each pilot firing two of them. This was followed by carrier qualifications aboard the USS Hornet (CVA-12) in December 1956 with both paddles and with the new mirror landing system.

During 1956, the squadron's Furies were involved in three major accidents. LT Daniel Weizer died on 17 April when he crashed short of the runway at Miramar after an engine oil pump failure. On 8 June LTJG Graber was killed during an ACM engagement with an F-100 when he crashed into a mountain 30 miles south of Riverside, CA. Another Navy versus Air Force engagement turned out badly for VF-142 on 9 October 1956. Two squadron pilots, ENS Rex Nelick

Above, VF-142 FJ-3M BuNo 139259. (William T. Larkins) Below, VF-142 FJ-3M BuNo 139252 being started aboard the USS Hornet (CVA-14). (via Mark Aldrich)

Above, VF-142 FJ-3M BuNo 141370. After the Hornet cruise, VF-142 received its new "NK" tail code. Trim was medium blue. (via Mark Aldrich) Bottom, VF-142 FJ-3M BuNo 139255. (William T. Larkins)

and LT Dick Powell collided at high altitude near Hemet, CA. Both pilots survived but not without difficulty. Ejection seat malfunctions affected both pilots. On the descent from 35,000 ft, Nelick finally was able to fire his seat at about 10,000; after numerous attempts. Powell's seat never did fire; he unstrapped while inverted and fell out of the aircraft.

CVG-14, VF-142 and the Hornet deployed to the Western Pacific from 21 January through 25 July 1957. In March, the squadron participated in Operation Beacon Hill and patrols with TF 77 off the Formosa Straits. Tragedy struck on 23 March when LTJG Jerry Jester stalled and spun-in on final 1/3 of a mile aft of the ship. He had been orbiting with a stuck open canopy and had become disoriented. On the 27th LT Ernie Genter was lost on a starboard catapult shot. In May, VF-142 participated in Operation True Blue which included four carriers and the Nationalist Chinese Air Force. During the cruise the squadron flew nearly 1,600 hours and made some 600 traps.

The Furies were replaced with Vought F8U-1 Crusaders in September 1957 and VF-142 received the F4H-1 in February 1962. The squadron was redesignated VF-96 on 1 June 1962 and the F4H-1 became the F-4B in September 1962. The squadron was disestablished on 30 June 1975.

FIGHTER SQUADRON ONE FORTY-THREE, VF-143 "KINGPINS"

Reserve squadron VF-821 was called to active duty on 20 July 1950 in response to the Korean War. CVG-101, with VF-821 attached, deployed to Korea aboard the USS Boxer (CV-21) from 27 March through 30 November 1951. The squadron returned to Korea aboard the USS Essex (CV-9) for their second Panther war cruise from 16 June 1952 through 6 February 1953 as part of ATG-2. On 4 February 1953, while still at sea, VF-821 was redesignated VF-143.

After arriving back in California, the squadron immediately began to transition to the F9F-6 Cougar. VF-143 returned to the Western Pacific on 1 December 1953 on board the Essex for a second tour with ATG-2. CVA-9 returned to CONUS on 12 July 1954. After a nine month turn-around period, VF-143 returned to WestPac aboard the USS Philippine Sea (CVA-47) from 1 April through 23 November 1955, again as part of ATG-2.

In late 1956 the squadron transitioned to the FJ-3M Fury. The squadron's previous name and patch, the "Cougars", was replaced with the "Kingpins". In keeping

Above, three VF-143 FJ-3s in flight in 1956. (USN) Below, VF-143 FJ-3 BuNo 139258. It was rare to ever see a Fury without wing tanks. (Ginter collection) Bottom, VF-143 FJ-3 BuNo 139229 in 1956. Trim, fin tip and wing tips were red. (William T. Larkins)

FIGHTER SQUADRON ONE FORTY-THREE, VF-143 "KINGPINS"

The FJ-3 skipper of VF-143, CDR Vince Kelly, renamed the squadron the Kingpins and became famous by flying formation at squadron strength in a ten-plane arrowhead formation that matched bowling pin set-up. Below, the cover of the November 1957 *Naval Aviation News*.

with their new name, VF-143's skipper CDR Vince Kelly ordered that formation flying at squadron strength would be done in a ten-plane arrowhead formation that

Above, VF-143 flightline during the 1957 Fleet Air Gunnery Meet. (USN) Below, BuNos 141433 and 141429 in flight in April 1957. (National Archives) Bottom, VF-143 FJ-3 BuNo 141406 landing at El Centro in April 1957. (USN)

matched a bowling pin set-up.

During a three-week El Centro gunnery deployment, the squadron maintained a 98.1% availabilty of aircraft and 25 gunnery "E"s were awarded. This was followed by a second-place win in the Pacific Fleet air-to-air gunnery meet and participation in the 1957 Naval Air Weapons Meet.

After training in the San Diego area the USS Hancock (CVA-19) and ATG-2 sailed north to NAS Alameda where all squadron equipment was loaded and a WestPac deployment commenced on 6 April 1957. The deployment ended on 18 September 1957. During the cruise the Kingpin pilots flew 1,919 flight hours and made 1,156 arrested landings.

VF-143 was disestablished on 1 April 1958.

Above right, VF-143 FJ-3M BuNo 141406 being re-spotted on the USS Hancock (CVA-19) in 1957. (USN) At right, VF-143 FJ-3M BuNo 141429 being pushed onto the elevator on the USS Hancock (CVA-19) in 1957. (USN) Below, VF-143 CAG bird FJ-3M BuNo 141411 traps aboard the Hancock. (USN)

FIGHTER SQUADRON ONE FIFTY-THREE, VF-153 "BLUE-TAIL-FLIES"

Reserve squadron VF-831 was called to active duty from NAS Floyd Bennett Field, NY, on 1 February 1951 with F6F Hellcats. The first Grumman F9F-2 arrived five days later to replace the obsolete Hellcats. They joined Air Group 15 at NAS Moffett Field, CA, in April then sailed aboard the USS Antietam (CV-36) for Korea on 8 September. By October the squadron was online with Task Force 77 off the coast of Korea. By May 1952, the squadron had returned to Moffett Field and received the F9F-5 Panther. On 24 January 1953, the squadron sailed for Korea aboard the USS Princeton (CVA-37). While at sea, the squadron was augmented into the regular Navy and redesignated VF-153 on 4 February 1953.

After returning to Moffett Field, the Panthers were replaced by F9F-6 Cougars in October 1953. In July 1954, VF-153 deployed aboard the USS Yorktown (CVA-10) with a nucleus of twelve Korean War Panther veterans.

The Cougars were replaced with FJ-3 Furies in March 1955. The Furies only lasted until November 1955 when they were replaced again with Cougars, this time the F9F-8. The F9F-8B started arriving to serve alongside the -8s in April 1956. On 15 December 1956, the squadron was redesignated VA-153 in preparation for receiving the Douglas A4D-1 Skyhawk on 12 February 1957.

The squadron went on to fly four other versions of the "Scooter": the A4D-2, A4D-2N/A-4C, A-4E, and A-4F. The Vought A-7A replaced the Skyhawks on 14 September 1969 and the A-7B was received in May 1973. VA-153 was disestablished on 30 September 1977.

Below, freshly-finished FJ-3 Fury being towed from the factory in 1955 is shown here as no VF-153 photos were discovered for use in this book. (via Mark Aldrich)

FIGHTER SQUADRON ONE FIFTY-FOUR, VF-154 "BLACK KNIGHTS"

The history of VF-837 can be traced back to reserve unit VBF-718 at Floyd Bennett Field, NAS New York. With F6F-5s the squadron was redesignated VF-68A on 1 February 1947. F4U-4 and FG-1D Corsairs were added and the squadron was reorganized under the Air Reserve program and designated VF-837 on 1 December 1949. Also in December the squadron received FH-1 Phantoms in addition to the Corsairs. They were called to active duty on 1 February 1951 because of the advent of the Korean War and were transferred to NAS Moffett Field, CA.

At the time of activation, the squadron was made up of 27 officers and 76 enlisted personnel. The squadron immediately started an extensive training program designed to speed conversion to the Grumman F9F-2 Panther and get them into the war.

Under the command of CDR J.J. Boydstun, VF-154 returned to NAS Moffett Field on 31 February 1955 and transferred to the FJ-3 Fury in July. The squadron missed its deployment with Air Group 15 aboard the USS Wasp (CVA-18) due to compressor problems with the FJ-3's engines. In October 1956, CDR F.X.

Above, officers of VF-154 in July 1956. Top row left-to-right: Cliff Thompson, Jim Ford, Lew Jesse, Paul Hamilton, and Ben Haverfield. Next row: Ted Martin, unknown, John Miotel, Burt Barclay, and Bob Lavender. Next row: unknown, Ed Shiver, unknown, John Howsen, Ray Rossi, Herb Trout. Bottom row: Bob Baldwin, Bob Pollard, Clancy Rich, CO Howard Boydston, Zeke Zelke, and unknown. (John Miottel) Below, VF-154 FJ-3 BuNo 135978 in 1956 belonged to LTJG E.C. Shiver. Lightning bolt, fuselage stripe and tail flashes were orange outlined in white. (William T. Larkins)

Timmes took command of VF-154.

In June 1957, the squadron transitioned to the F8U-1 Crusader. The F-4B replaced the Crusader in April 1966 and the squadron moved up to F-4Js and finally F-4Ns before the Phantoms were replaced by F-14A Tomcats in May 1984.

Above, LT Shiver's mount in 1957, BuNo 135928, with orange trim. (William T. Larkins) Below, October 1956 change of command ceremony at NAS Moffett Field, CA. (John Miottel)

FIGHTER SQUADRON ONE SEVENTY-THREE, VF-173 "JESTERS"

Above, VF-173 FJ-3 positioned on the Bennington's catapult in 1954. (USN/Tailhook) Below, VF-173 FJ-3 with original "R" tail code conducting FCLP practice at NAS Miramar, CA, in 1954. (via Tailhook)

Fighter Squadron One Hundred Seventy-Three (VF-173) was established on 11 August 1948. During the Korean War, the unit flew the Chance Vought F4U Corsair. After returning from a Mediterranean deployment, VF-173 transitioned to the F9F-6 Cougar in early 1953.

The squadron deployed aboard the USS Wasp (CVA-18) from September 1953 through April 1954 for a post-Korea cruise. The deployment found the Wasp transiting from the East Coast to the West Coast through the Mediterranean, Suez Canal, Colombia, Manila, and on to Yokosuka. The ship joined Task Force 77 for peacekeeping operations off Korea. The primary mission of VF-173 was combat air patrol (CAP). After the cruise, the ship reported to NAS Alameda and VF-173 transitioned to the FJ-3 Fury.

Carrier qualifications were conducted by the squadron in 1954-1955 aboard the USS Benington (CVA-20). This was followed by a deployment aboard the USS Intrepid (CVA-11) from 24 May through 24 November 1955. During carrier qualifications aboard the USS Randolph (CVA-15) on 24 June 1957, LTJG Don Mairose took the barrier in BuNo 139272 and struck a pack of ADs and F-2H-3s causing minimal damage. In April 1957 while aboard the USS Franklin D. Roosevelt (CVA-42), VF-173 took part in cold weather exercises in the North Atlantic. A Med cruise was then conducted from 8 June 1957 through 23 February 1958 aboard CVA-42 with FJ-3Ms.

VF-173 was disestablished on 1 February 1959.

VF-173

Above, VF-173 FJ-3 crosses the ramp on CVA-20 in 1954. (Emery Vrana via Tailhook) At right, loading VF-173 FJ-3 BuNo 135798 aboard CVA-20 in 1955. (via Tailhook) Below, VF-173 FJ-3 F/319 on the deck edge elevator aboard the USS Bennington. (Ginter collection)

Above, three VF-173 FJ-3s F/308, F/309 and F/313 being inspected for launch from CVA-20 with the pilots waiting in the cockpit. (Ginter collection) Below, CO's aircraft tensioned for launch with LTJG R.B. Donnelly awaiting his turn. (Ginter collection)

VF-173 FJ-3s share the deck aboard the USS F.D.Roosevelt (CVA-42) with F7U-3 during cold weather exercise in April 1957. Tail code was "R". (National Archives) Below, deck handling during cold weather operations in the North Atlantic. (National Archives)

Above, four VF-173 FJ-3M Furies over the Med in 1958. BuNos 139261 AD/204, 139274 AD/209, 139270 AD/205, and 139271 AD/206. (USN) Below, VF-173 FJ-3M with Sidewinder rails aboard CVA-42 in 1958. Trim colors were yellow. (USN)

VF-173 FJ-3 BuNo 139247 with an "R" tail code. The squadron wore three tail codes; "F" 1954-56, "R" 1956-57 and "AD" 1957-58. (via John Elliott)

FIGHTER SQUADRON ONE SEVENTY-FOUR, VF-174 "HELLRAZORS"

The Hellrazors were originally established as Bombing Squadron Eighty-One (VB-81) on 1 March 1944. The squadron flew various versions of the Curtiss Helldiver and was redesignated Attack Squadron Thirteen A (VA-13A) on 1 March 1944. On 2 August 1948, the unit became VA-134 and was flying the F4U-4 Corsair when it was redesignated Fighter Squadron One Hundred Seventy-Four (VF-174) on 15 February 1950. They received the F4U-5 in June 1950. In January 1953, the squadron entered the jet age when it received the F9F-6 Cougar.

As part of CVG-1, VF-174 deployed aboard the USS Midway (CVA-41) during its world cruise from 27 December 1954 through 14 July 1955. In September, the squadron transitioned to the FJ-3 Fury which it flew until March 1956 when the F9F-8 Cougar was acquired.

Crusaders started arriving in October 1957 and on 28 April 1958 the squadron's mission was changed to the East Coast F-8 training squadron. TV-2s and F9F-8Ts were also acquired for this mission which VF-174 flew until 14 October 1966 when they transitioned to the A-7A Corsair. VF-174 was redesignated VA-174 on 1 July 1966 and went on to fly all versions of the Corsair II before its disestablished on 30 June 1988.

VF-174 FJ-3s in formation in January 1956 with BuNo 136050 (T/202) in the foreground. Aircraft trim was orange/yellow. (NMNA)

FIGHTER SQUADRON ONE NINETY-ONE, VF-191 "SATAN'S ANGELS"

VF-191 was originally established as VF-19 at NAAS Los Alamitos, CA, on 15 August 1943. VF-19 flew Hellcats during World War Two and became the first squadron to receive the new F8F-1 Bearcat. On 15 November 1946, the squadron was redesignated VF-19A and then VF-191 on 24 August 1948. In 1949, the squadron transitioned to the Grumman F9F-2B Panther.

After the Korean War started, the Navy's Flight Demonstration Team (Blue Angels) was disbanded and the skipper along with six pilots and thirty-three enlisted men were transferred to VF-191 where they provided the very experienced nucleus for the "Kittens" first war cruise.

The squadron transitioned to the F9F-6 Cougar in 1953. In October 1955, CDR Jim Swope, a WWII double ace, assumed command and in 1956 the FJ-3 Fury arrived

At left, CDR Swope was CO of VF-191 while equipped with the FJ-3. He is seen here aboard the Yorktown. (USN) At top, FJ-3M BuNo 136028 over the Pacific in June 1957. (USN) Above, VF-191 Fury escorts a VAW-11 AD-2Q past the plane guard. (USN) Below, echelon formation of four VF-191 FJ-3s BuNos 135828 B/113, 136038 B/110, 136033 B/109, and 135936 B/104 on 17 June 1957. (USN)

VF-191 deployed with their FJ-3s

aboard the USS Yorktown (CVA-10) as part of CVG-19 from 9 March through 25 August 1957. During the cruise, squadron pilots set a Pacific Fleet record of 1,317 accident-free carrier landings.

In October 1957, the F11F-1 Tiger began replacing the squadron's Furies. In June 1960, the squadron received the F8U-1 Crusader and transitioned to the F-4J Phantom II in March 1976. VF-191 was disestablished on 1 March 1978 at NAS Miramar, CA.

At right, VF-191 FJ-3M being hooked-up for a catapult shot off the Yorktown. (USN) Below, four VF-191 Furies in flight in 1957. (USN)

Above, BuNo 135828 taxis on Yorktown in 1957. VF-191's Furies were trimmed in red. (NMNA) Below, BuNos 136049 B/111 and 136079 B/115 prepare for launch from Yorktown. (NMNA)

At right, "Pilots man your planes!" VF-191 pilots race for their aircraft during a drill on Yorktown in 1957. (USN) Below, VF-191 FJ-3s are deck background during a ceremony aboard the USS Yorktown (CVA-10) in June 1957. (National Archives)

FIGHTER SQUADRON TWO - ELEVEN, VF-211 "RED CHECKERTAILS"

Fighting Squadron Two-Eleven was established with the FJ-3M at NAS Moffett Field, CA, in June 1955 and was assigned to CVG-21. The unit participated in the Fleet Air Gunnery Meet at NAAS El Centro in April 1956.

VF-211 completed one deployment with their FJ-3Ms. It was aboard the USS Bon Homme Richard (CVA-31) for a West Pac cruise from 16 August 1956 through 28 February 1957. This was the first cruise of CVA-31 after its modernization which included a new angled deck.

After returning to CONUS, CDR D.C. Davis relieved CDR Pugh and in March 1957 the squadron began transition to the F8U-1 Crusader. The squadron deployed aboard the USS Midway (CVA-41) in August 1958 and on 9 March 1959 was redesignated VF-24 when it was reassigned to CVG-2. The squadron was redesignated VF-214 from 1 September to 17 September 1964 before reverting back to VF-24. In the winter of 1975, the squadron transitioned to the F-14A Tomcat.

At top right, VF-211 FJ-3 in flight near Japan in February 1957. (USN) At right, four VF-211 FJ-3s (G/105, G/106, G/107, G/110) in flight on 11 February 1957. (National Archives)

VF-211 "RED CHECKERTAILS"

Above, VF-211 flightline at NAAS El Centro, CA, during the Fleet Air Gunnery Meet in April 1956. (NMNA) At left, VF-211 FJ-3 during carrier qualifications aboard the USS Bon Homme Richard (CVA-31) in June 1956. (USN) Below, VF-211 pilots pose in front of a squadron Fury and behind a Fighting 211 sign. (via Mark Aldrich) At right, VF-211 FJ-3s near Mt. Fuji on 11 February 1957. (National Archives) Below right, Sidewinder-armed VF-211 FJ-3M prepares to launch from the USS Bon Homme Richard (CVA-31) in June 1956. Aircraft trim was red. (National Archives)

G
110
NAVY
105
NAVY
G
NAVY
G
NAVY
G

113
DANGER
JET INTAKE
13
G
NAVY

Above, eleven VF-211 Furies spotted on the afterdeck of the USS Bon Homme Richard in 1956. (USN) Below, VF-211 Red Checkertail FJ-3M catapults off CVA-31 during carrier qualifications in 1956. (USN)

FIGHTER SQUADRON TWO - FOURTEEN, VF-214 “VOLUNTEERS”

Above, acting CO LCDR R.M. Soule delivering the squadron’s first aircraft, an FJ-3 Fury, to NAS Moffett Field, CA, in March 1955. (USN) Below, illustration of squadron aircraft. Trim was medium blue outlined in white. (courtesy Squadron Signal publications)

Fighter Squadron Two Hundred Fourteen (VF-214) was officially established on 30 March 1955 at NAS Moffett Field, CA. Earlier that month, the squadron’s first aircraft, a North American FJ-3 Fury, was delivered by the unit’s acting Commanding Officer, LCDR R.M. Soule. Soule, who was relived by CDR I.T. Raynor on 7 June 1955, became the squadron’s Executive Officer.

The Furies were replaced with Grumman F9F-8 Cougars in November 1955. As part of ATG-4, VF-214 deployed aboard the USS Yorktown (CVA-10) to the Western Pacific from 19 March through 15 September 1956. VF-214 was redesignated Attack Squadron Two Hundred Fourteen (VA-214) on 11 October 1956.

On 18 January 1958, the squadron traded-in their Cougars for North American FJ-4B Furies. VA-214 became the first squadron on the West Coast to make an extended deployment with the new FJ-4B Fury. The squadron deployed aboard the USS Hornet (CVA-12) from 18 January through 2 July 1958 as part of ATG-4. The squadron was disestablished on 1 August 1958.

VF-214 profile courtesy of Squadron Signal Publications. Fuselage trim was light blue bordered by white.

MARS-27

Above right, H&MS-31 FJ-3 BuNo 136052 at MCAS Miami, FL, in 1958. Drop tanks were da-glo red. (via Clay Jansson) Below right, H&MS-32 FJ-3 BuNo 136012 operated by the Commander of MAG-32, COL R.H. Elwood, in May 1958 at Andrews AFB, MD. Trim was blue and black. (Ginter collection)

H&MS-32

MARINE UTILITY SQUADRONS

Above, FJ-3 assigned to MARS-27 in 1956. (NMNA)

H&MS-31

Headquarters Squadron 31 was established on 1 February 1943 at MCAS Cherry Point. Following WWII, H&MS-31 was disestablished on 31 May 1947. Due to the Korean War, H&MS-31 was reactivated on 17 March 1952 at MCAS Cherry Point again. In the late '50s they flew the FJ-3/3M before once again being disestablished on 31 January 1959. In 1961, H&MS-31 was reactivated a third time. This time they were assigned to MCAS Beaufort.

Like H&MS-31, Headquarters Squadron 32 was established at Cherry Point, NC, on 1 February 1943. After WWII, H&MS-32 was deactivated at MCAS El Toro on 30 April 1947. Because of the Korean War, the squadron was reactivated at MCAS Miami, FL, on 8 May 1952. While flying FJ-3/3Ms, the squadron operated from MCAS Cherry Point and MCAS Beaufort beginning in August 1957. The squadron was redesignated MALS-32 in October 1988.

MARINE FIGHTER SQUADRON ONE TWENTY-TWO, VMF-122 "CANDYSTRIPERS"

VMF-122 was established on 1 March 1942 at Camp Kearney, CA. Initially equipped with F4F Wildcats, the squadron finished the war with F4U Corsairs but stayed in the Pacific on the island of Peleliu until 1946. They were transferred to MCAS Cherry Point, NC, and deactivated in July 1946.

The squadron was re-established in October 1946 with F4U-4 Corsairs and became the first operational Marine jet squadron when it acquired the FH-1 Phantom in November 1947. Under the command of LTCOL Marion E. Carl, the squadron formed an all-jet flight demonstration team around the Phantom in 1948. The aerial demonstration team was disbanded in 1949 when VMF-122 received the F2H-2 Banshee.

The squadron was scheduled to receive F9F-4 Panthers in the spring of 1952 and had transferred-out most of their Banshees when a grounding order halted the Panthers' delivery. In the interim, ten F6F-5 Hellcats were acquired. The F9F-4 began arriving in July, but was replaced within a few months with the F9F-5.

In January 1954, the squadron started receiving the FJ-2 Fury and became known as the Candystripers. Late in 1955 they transitioned to the FJ-3 Fury and later the FJ-3M. In 1956-57, the squadron made several short training deployments aboard the USS Forrestal and USS Saratoga.

In December 1957, VMF-122 became the first Marine squadron to equip with the F8U-1 Crusader. The F-4B Phantom arrived on 1 July 1965 and the F/A-18 Hornet arrived in 1986.

Below, four VMF-122 FJ-3s in flight on 3 April 1956. They were blue with red and gold candystripes. (USMC)

Above, VMF-122 FJ-3 BuNo 139211 on 30 July 1956 repainted in gull grey and white. (USMC) Below, 139211 in flight with BuNo 136134 on 30 July 1956. Aircraft had three red stripes and two gold stripes on the tail fin. (USMC)

MARINE FIGHTER SQUADRON TWO THIRTY-FIVE, VMF-235 "DEATH ANGELS"

VMSB-235 was established on 1 January 1943 at MCAAS El Centro, CA. After participating in five campaigns during World War Two, the squadron was deactivated on 10 November 1944.

The Death Angels were reactivated as VMF-235, a reserve squadron, on 15 November 1946. They were recalled to active duty from NAS Squantum, MA, with F4U-4 Corsairs and transferred to MCAS El Toro, CA, on 12 August 1950. In September 1952, they transitioned to the Grumman F9F-2 Panther.

The squadron transitioned to the FJ-2 Fury in March 1954 and the FJ-4 Fury in 1956. VMF-235 deployed its FJ-4s to Japan in 1956-57. On their return to CONUS in 1957, the squadron reformed at Parris Island, SC. They had left their FJ-4s in Japan and received FJ-3Ms in South Carolina. The Furies were operated into late 1958 when they were replaced with the F8U-1 Crusader.

Above, FJ-3M BuNo 141393. (via W.T Larkins) Below, VMF-235 FJ-3M DB/16 at South Carolina. Aircraft trim was red with white stars. (USMC)

. The squadron was redesignated Marine Fighter (All Weather) Squadron 235 (VMF(AW)-235) in February 1962 while flying the F-8D. In September 1968, the squadron was redesignated Marine Fighter

Attack Squadron 235 (VMFA-235) and equipped with the F-4J Phantom II. F/A-18C Hornets replaced the Phantoms in November 1989 and the squadron was disestablished on 14 June 1996.

VMF-235 line at Parris Island on 21 May 1958. (USMC) Below, flight of four VMF-235 FJ-3Ms on 27 September 1957. (via Mark Aldrich)

MARINE FIGHTER SQUADRON TWO FIFTY-ONE, VMF-251 "THUNDERBOLTS"

VMF-251 was established at North Island, CA, as VMO-251 in 1944 flying the F4F Wildcat. They switched to the F4U Corsair and were redesignated VMF-251 in December 1944. The squadron was disestablished on 1 June 1945 only to be reestablished on 1 July 1946 as a reserve unit at NAS Grosse Isle, MI.

Due to the Korean War, the squadron was called to active duty on 1 March 1951 and became VMA-251. The squadron reported to MCAS Miami, FL, on 20 April 1957 where they received the North American FJ-3 Fury and became once again VMF-251.

In late 1958, the squadron reported to MCAS El Toro, CA. where they transitioned to the Vought F8U Crusader. In October 1964, they transitioned to the F-4B Phantom and were redesignated VMFA-251. The F/A-18A replaced the Phantom in 1986.

Above, VMF-251 squadron personnel pose with their FJ-3s at NAF Roosevelt Roads, PR, in March 1958. (USMC) Below, VMF-251 FJ-3 BuNo 135935 was adorned by a white lightning bolt bordered by dark blue on top and red below. Tail stripe is believed to be orange bordered by dark blue. (Warren Ship via Tailhook)

MARINE FIGHTER SQUADRON THREE TWELVE, VMF-312 "CHECKERBOARDS"

Above, VMF-312 FJ-3M catches the "Morset" cable at Coco Solo, Canal Zone in February 1957. (USMC) Below, LCOL Smunk fired the 100,000th 20mm round from an FJ-3M. (USMC)

VMF-312 was established at Parris Island, S.C., on 1 June 1943. On 1 March 1952, while deployed to Korea with F4U Corsairs, VMF-312 was redesignated VMA-312.

After the Korean War, while at Itami, Japan, the squadron was reduced in personnel to Capt. Harry B. Stuckey, who brought the Checkerboards' flag to MCAS Miami, FL, on 16 June 1953. On the 21st, VMA-312 was assigned to MAG-32, 3rd MAW, and three days later, Maj. Charles A. House assumed command. The reformed squadron began receiving Grumman F9F-4 Panthers and was transferred to MCAS Cherry Point, N.C., on 5 January 1954. On 15 February, the Checkerboards were redesignated VMF-312 and received FJ-2 Furies in November 1954. The FJ-2 was replaced by the FJ-3 in October 1956 which were supplemented with FJ-3Ms in November 1958.

In February 1957, the squadron made a short deployment to the Canal Zone and carrier qualified aboard the USS Saratoga (CVA-60) in July. VMF-312 made the Marines only FJ-3/3M carrier deployment. They sailed aboard the USS Lake Champlain (CVS-39) on a short Mediterranean cruise from 5 September through 31 October 1957 as part of the all-Marine Air Group MAG-26. The squadron shared the deck with VMA-533 (F2H-4), VMA-324 (AD-6), HMR(L)-261 (HUS-1), HMR(L)-262 (HRS-3), HMR(M)-461 (HR2S-1), and HU-2 DET 34 (HUP-2). Once in the Med they participated in joint operations with the Hellenic Air Force from Tanagra. MAG-26 also participated in NATO exercise Deepwater, which supported amphibious landings on the Turkish coast. This was followed by flood relief to Valencia, Spain, on 16 October.

On 13 November 1957, the squadron relocated to MCAS Beaufort, SC. The Commanding Officer, LCOL Thomas H. Hughes, was lost in March 1958 during an overwater calibration flight. From October through December 1958, VMF-312 was stationed at Guantanamo Bay, Cuba, to protect the base during the Castro-led revolution. Gunnery training was conducted

almost daily with the squadron CO, LCOL Louis R. Smunk firing the 100,000th 20mm round of their Gitmo deployment.

The squadron went supersonic in June 1959 with the arrival of the F8U-1 Crusader. On 1 August 1963, VMF-312 became Marine All Weather Fighter Squadron 312. VMF(AW)-312 was redesignated Marine Fighter Attack Squadron 312 on 1 February 1966 when the F-4B Phantom was received. F/A-18A Hornets replaced

At top, two VMF-312 Furies simulate close air support for Marines charging ashore at Camp Lejeune, NC, on 27 January 1957. (USMC) Above, FJ-3M BuNo 139272. (USMC) Below, loading VMF-312 FJ-3M BuNo 135180 onto the USS Lake Champlain (CVA-39). (via Fred Roos)

MARINE FIGHTER SQUADRON THREE THIRTY-THREE, VMF-333
"FIGHTING SHAMROCKS"

VMSB-333 was established at MCAS Cherry Point on 1 August 1943 with the SBD-4 Dauntless. The squadron moved to MCAAF Bogue on 6 September and deployed to NAS Bocca Chica in April 1944 for ASW training. By July 1944 Trip Trey was operating from Midway Island. On 14 October the squadron was redesignated VMBF-333 and transitioned to the F4U-1D Corsair. On 30 December they reverted to VMSB-333 and were deactivated at Ewa on 1 November 1945.

The squadron was reestablished on 1 August 1952 at MCAS Miami as VMA-33. Initially equipped with the F6F-5 Hellcat, they transitioned to the F4U-4 Corsair a few months later. In February 1954, the squadron equipped with AD-5 Skyraiders and also operated AD-4s and AD-6s until January 1957, when they received their first FJ-3 Fury and were redesignated VMF-333.

In January 1958, the squadron boarded the USS Forrestal (CVA-59) for carrier qualifications and participation in Operation Lant Phibex 1-58. 300 sorties and 350 flight hours were flown during this short cruise. The at-sea period was followed by a transfer to MCAS Beaufort, SC.

The squadron conducted the first evaluation of the new tactical landing system in November 1958 and by February 1959 all pilots had completed Phase One training. The FJ-3/3Ms gave way to the F8U-2 Crusader starting on 6 November 1959.

The squadron's Crusaders were traded-in for the F-4J Phantom in 1968.

Below, VMF-333 flightline with its green and white-trimmed FJ-3Ms. (USMC) Bottom, VMF-333 FJ-3 BuNo 136022. (Clay Jansson)

MARINE FIGHTER SQUADRON THREE THIRTY-FOUR, VMF-334 "FALCONS"

The Falcons were established at MCAS Cherry Point on 1 August 1943 as Marine Scout Bombing Squadron VMSB-334. They flew the SBD Dauntless until they were deactivated on 10 October 1944.

On 31 May 1952, the squadron was reactivated at MCAS Miami, FL, as VMF-334, and shortly thereafter became VMA-334. The unit's authorized strength was 53 aviators, 7 ground officers, 152 enlisted men and 24 aircraft. It was not until August of 1952 that the squadron received enough F6F-5s to conduct normal operations. A few months later the F4U-4 Corsair started replacing the Hellcats and the squadron started receiving Grumman F9F-4 Panthers on 1 May 1953.

On 23 January 1954, the squadron was transferred to MCAS Cherry Point, NC, and was assigned to MAG-32. The squadron was redesignated VMF-334 again on 1 July 1954 and the Panthers were replaced by the North American FJ-2 Fury in the summer of 1955. The North American FJ-3M Fury replaced the FJ-2 Furies in 1956, and in January 1957 the North American FJ-4 Fury replaced the missile-equipped FJ-3Ms.

The supersonic Crusader arrived in January 1958 and the squadron was redesignated VMF(AW)-314 on 15 February 1958. In June 1967, they became the first Marine squadron to receive the McDonnell F-4J Phantom II and were redesignated Marine Fighter Attack Squadron 334 (VMFA-334). VMFA-334 was disestablished at MCAS El Toro, CA, on 30 December 1971.

Above, VMF-334 FJ-3M BuNo 141374 in 1956. (Ginter collection) Below, the Falcons' flightline at Cherry Point, NC, on 28 June 1956. VMF-334 FJ-3Ms had a "MX" tail code painted on a blue tail stripe. The Falcon was orange. (National Archives via Fred Roos)

NARTU MIAMI, FLORIDA "6H"

At right, white and orange Miami Reserve FJ-3 BuNo 136001 in October 1963. (Dusty Carter)

NAS COLUMBUS, OHIO "7C"

At left, retired NAS Columbus Reserve FJ-3M BuNo 135857 at NAF Litchfield Park, AZ, on 21 March 1960. Aircraft was gull grey and white with an international orange fuselage band. (William Swisher)

NAS DALLAS, TEXAS "7D"

At right, flight of four NAS Dallas, TX, Reserve FJ-3/3Ms in the gull grey, white and da-glo red paint scheme. (Ginter collection)

Above, retired NAS Dallas white and orange FJ-3M BuNo 139221 at NAF Litchfield Park on 21 March 1960. (William Swisher) Below, four Dallas-based FJ-3Ms in flight. Markings were unusual in that aircraft numbers and tail and wing codes were made up with a series of dashes. Aircraft were gull grey and white with an international orange fuselage stripe. (via Tailhook)

NAS NEW YORK, “7R”

At top, gull grey and white FJ-3 BuNo 135998 on 17 May 1958 at Floyd Bennett Field. (via Lionel Paul) Above, NAS New York received a number of FJ-3D drone control aircraft of which BuNo 136038 was one. (7-10-59 via Lionel Paul) Below, also on 10 July 1959 BuNo 135820 was seen in a gull grey and white scheme with an international orange fuselage stripe. (Abbott Hafter) Bottom, FJ-3 BuNo 135832 in white and orange colors on 13 September 1959. (Abbott Hafter)

NAS WILLOW GROVE, "7W"

At top, white and international orange Willow Grove-based FJ-3 BuNo 135902 in flight in the spring of 1959. (Robert Hogan) Above, BuNo 135915 was a gull grey and white Willow Grove-based FJ-3 with international orange fuselage stripe above and below the Navy-Marine on the aft fuselage on 3 September 1959. The Fury has a blue drop tank on the right wing and a gull grey and white one on the left wing. (Roger Besecker) Bottom, Willow Grove-based FJ-3 BuNo 136039 in 1961. Speed brake interior was red. (Robert O'Dell via Norm Taylor)

Above, Willow Grove VF-931/932 FJ-3s BuNos 135935 7W/10 and 135865 7W/14 in April 1961. Aircraft were gull grey with da-glo red-orange nose, tail, wing tanks and lengthwise wing stripes. (Robert Hogan) Below VF-931/932 white and international orange Willow Grove-based FJ-3 BuNo 136106 in the spring of 1959. (Robert Hogan)

NAS GLYNCO, "4B"

NAS SOUTH WEYMOUTH, "7Z"

Above, F-1C BuNo 135935 was assigned to the CIC school at NAS Glynco, GA, on 2 April 1966. Aircraft was white with deep red trim. Upper portion of wing tanks was black. (via Mark Aldrich) Below, FJ-3 BuNo 139251 in the blimp hangar at NAS South Weymouth during an open house in May 1960. Aircraft was gull grey and white with da-glo red nose, tail and outer wing panels. (Jim Burridge) Bottom, South Weymouth-based FJ-3 BuNo 135850 was gull grey with a white rudder and da-glo red-orange nose, tail, wing tanks and lengthwise upper wing stripes. (William T. Larkins)

Above, BuNo 136022 finished as a VMF-323 aircraft, however the squadron never flew the FJ-3/3M. (via Norm Taylor) Below, BuNo 136022 finished as a VMF-333 aircraft in 1987. (Ginter) Bottom, BuNo 136022 once again in bogus VMF-323 markings after its move from El Toro to Miramar. (Ginter) At top right, BuNo 141393 at Stockton, CA, with VMF-235 on the fuselage side but with a VMF-232 "WT" tail code. (William T. Larkins) At right, BuNo 135868 finished as a VF-191 Fury aboard the USS Intrepid (CVA-11) Museum in New York harbor in November 1988. (via Norm Taylor) Bottom right, Chino's Planes of Fame FJ-3M finished as BuNo 141435 in the markings of VF-73 in May 1989. Aircraft is actually BuNo 135867. (Nick Williams)

WT
1393
24
MARINES
FJ-3
141393
VMF-235

01
B
5868
NAVY
135868
VF-191
101

313
NAVY
VF-73
13

1956 BLUE ANGELS EVALUATION

Above, CDR Cormier leads the Blue Angels in flight while evaluating four FJ-3s (BuNos 141364, 141371, 141375, and 141379) at Columbus, OH, in 1956. (NAA) Below, Blue Angels inspecting an FJ-3 which they evaluated as a replacement for the the F9F-8 along with the A4D before deciding on the F11F Tiger. CDR "Zeke" Cormier is sitting in the cockpit. (via Mark Aldrich)

ESCI 1/48 SCALE FJ-3 FURY

The very old ESCI kit is not recommended. At best its a bad character of the real thing. If you carved a new fuselage and created an accurate canopy you could then make due with the kits wings, tailhook, landing gear and other Navy specific parts.

Below right, ESCI kit built by Brandon Stewart and pictured from the only angle that minimizes it's inaccuracy.

FALCON 1/72 SCALE FJ-3 CONVERSION

Falcon triple conversion kit IX (#4605) has been around over 20 years. It included the C-2, AD-5W and FJ-3. The kit has 12 high quality vacuform parts and a crystal clear canopy with instructions. You must provide a Heller F-86F for it;s wings and scratch build from the spares box the landing gear, wheels, cockpit interior, seat and decals. Two fuselage inaccuracy's are the instrument panel coaming and the lack of rudder ribbing. It is recommended that the panels be rescribed on both the vacuform parts and the Heller wings. The gun throughs should be recut too. **By Louis Santos**

COLLECT AIRE 1/48 SCALE FJ-3 FURY

The resin Collect Aire kit includes metal parts, decals, and 3 vacuform canopies. At first glance I was impressed with the general quality of the castings even though my copy had a concentration of bubbles around the wing fences. The engineering is very good by today's standards. It features a deep intake trunk with an engine front, wheel and gun bays and a deep exhaust pipe. The general outline is excellent as is the exterior paneling. The metal parts were acceptable except for the canopy frame which required a lot of filling and sanding. The vacuform canopies were unacceptable due to imperfections and I ended up vacuforming a new one. Choice of 2 markings are included on the excellent decal sheet. Also, a couple of ports and a cooling intake must be added to the fuselage. **By Louis Santos**

Below, Collect Aire FJ-3M built by Ron Dozbrelicky in VF-173 markings. (Courtesy of Collect Aire) Bottom, Collect Aire Fury built by Louis Santos with highly detailed enhancements including open oxygen compartment, ammo bay door, speed brake and gunbay.

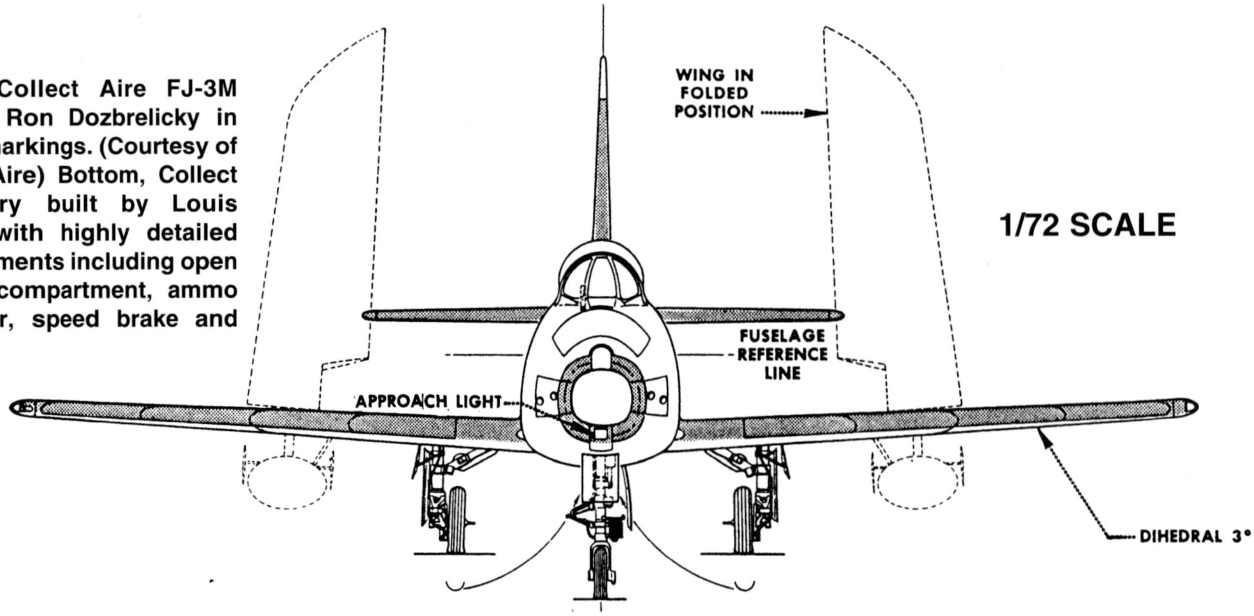